T0261703

FUNDAMENTALS OF
MICROBIOME SCIENCE

Fundamentals of Microbiome Science

How Microbes Shape Animal Biology

Angela E. Douglas

PRINCETON UNIVERSITY PRESS

PRINCETON AND OXFORD

Published by Princeton University Press,
41 William Street, Princeton, New Jersey 08540

In the United Kingdom: Princeton University Press,
Oxford Street, Woodstock, Oxfordshire OX20 1TR

press.princeton.edu

Cover illustration: Nanoflight SEM of the microbiome of the sponge *Aplysina aerophoba*.
Courtesy of Stefan Diller (University of Wurzburg) and Ute Hentschel-Humeida
(GEOMAR Helmholtz Center for Ocean Research, Kiel)

Library of Congress Cataloging-in-Publication Data

Names: Douglas, A. E. (Angela Elizabeth), 1956– author.
Title: Fundamentals of microbiome science : how microbes shape animal biology/ Angela E. Douglas.
Description: Princeton, New Jersey : Princeton University Press, [2018] |
    Includes bibliographical references and index.
Identifiers: LCCN 2017021144 | ISBN 9780691160344 (hardcover : alk. paper)
Subjects: | MESH: Microbiota | Microbial Interactions | Biological Phenomena | Animals
Classification: LCC QR1 | NLM QW 4 | DDC 571.6/3829--dc23 LC record available at
    https://lccn.loc.gov/2017021144

British Library Cataloging-in-Publication Data is available

This book has been composed in Adobe Text Pro and Gotham

Printed on acid-free paper. ∞

Printed in the United States of America

10 9 8 7 6 5 4 3 2 1

CONTENTS

Preface   vii

1   Introduction: Animals in a Microbial World   1

1.1. What Is An Animal?   1

1.2. Terminology: Dismantling the Tower of Babel   3

1.3. The Microbiology of Animals   5

1.4. Scope of this Book   9

2   The Ancient Roots of the Animal Microbiomes   12

2.1. Introduction   12

2.2. The Social Life of Bacteria   13

2.3. The Multiorganismal Origins of Eukaryotes   20

2.4. The Ubiquity of Microbial Associations in Eukaryotes   23

2.5. The Animal Condition   29

2.6. Summary   36

3   The Microbiome and Human Health   38

3.1. Introduction   38

3.2. The Biogeography of the Human Microbiome   39

3.3. How to Study Microbiota Effects on Human Health   45

3.4. The Microbiota and Human Disease   55

3.5. The Mass Extinction Event Within?   60

3.6. Summary   64

4   Defining the Rules of Engagement:
The Microbiome and the Animal Immune System   66

4.1. Introduction   66

4.2. Immune Effectors and the Regulation of the Microbiota   68

4.3. The Effects of the Microbiota on Animal Immune Function   76

*4.4. Symbiont-Mediated Protection: Microbiota as the
Second Immune System*   84

4.5. Summary   91

5 Microbial Drivers of Animal Behavior   93

  5.1. Introduction   93

  5.2. Microbes and Animal Feeding Behavior   95

  5.3. Microbial Arbiters of Mental Well-Being   102

  5.4. Microbes and Animal Communication   112

  5.5. Summary   118

6 The Inner Ecosystem of Animals   121

  6.1. Introduction   121

  6.2. The Abundance and Distribution of Animal-Associated Microorganisms   123

  6.3. Ecological Processes Shaping Microbial Communities in Animals   127

  6.4. Functions of the Inner Ecosystem   137

  6.5. Summary   150

7 Evolutionary Processes and Consequences   152

  7.1. Introduction   152

  7.2. Costs and Benefits   153

  7.3. Evolutionary Specialization and Its Consequences   164

  7.4. Symbiosis as the Evolutionary Engine of Diversification   173

  7.5. Summary   189

8 The Animal Reimagined   191

  8.1. Introduction   191

  8.2. The Scope of the Animal   192

  8.3. The Determinants of Animal Phenotype   197

  8.4. The Animal in the Anthropocene   200

References   207

Index   231

PREFACE

This book is about animal microbiomes: the microorganisms that inhabit the body of animals, including humans, and keep their animal hosts healthy. In recent years, animal microbiomes have become a hot topic in the life sciences. Academic, commercial, and funding institutions are investing in major microbiome research centers and funding initiatives; microbiomes are the topic of special issues in journals, conference symposia, and new undergraduate and graduate courses; microbiomes have twice been a *Science* journal "breakthrough of the year" (in 2011 and 2013); and the US National Microbiome Initiative was announced from the White House in May 2016.

Why all the excitement about microbiomes? The reasons are twofold: microbiome science provides a radically different way to understand animals, and it offers the promise of novel therapies, especially for human health.

And why have I written this book? My purpose is to communicate that we can only understand fully how animals function by appreciating that an animal interacts unceasingly with the communities of microorganisms that live within and on the surface of its body. This book provides the evidence for these interactions, and introduces the main concepts that shape how the discipline of microbiome science is conducted. Some of the concepts are new, for example driven by recent developments in genome science, but others have been developed over many decades of research in the related discipline of symbiosis. The two disciplines of microbiome science and symbiosis overlap extensively, and although this disciplinary ambiguity may appear untidy, it enriches our understanding of the microbiology of animals.

There is a third question: Who is this book for? It is for everyone with tertiary level education in biology who is curious about microbiomes. This book complements the recent excellent books that do not require advanced biological education, such as Ed Yong's *I Contain Multitudes* and Martin Blaser's *Missing Microbes*. *The Animal Microbiome* aims to be useful to researchers, to colleagues developing courses on microbiomes, and to students. In particular, I hope that at least a few of the microbiome gainsayers who declare that microbiome science "is all correlation with no substance" will read this book, and be persuaded by the mounting evidence for the pervasive effects of the microbiome on animal biology. Equally, I trust that the "microbiome enthusiast" who is tempted to attribute any animal trait with an unknown cause to the microbiome will appreciate the evidence that animal-microbiome interactions are inherently complex and frequently context-dependent.

My decision to write this book and the way it is structured are the consequence of countless conversations with colleagues, students, and friends over the last decade. I thank you all. I have special thanks to colleagues who read individual chapters: Karen Adair, Nana Ankrah, Anurag Agrawal, Steve Ellner, Wolfgang Kunze, Brian Lazzaro, and Karl Niklas. Their thoughtful comments and corrections have greatly improved the book. I am also grateful to Alison Kalett, the editor at Princeton University Press, for her support and good advice. Finally, my thanks as always to Jeremy for his consistent encouragement and steadfast belief that this book really would get written.

Angela Douglas
1 October 2016

FUNDAMENTALS OF MICROBIOME SCIENCE

1

Introduction

ANIMALS IN A MICROBIAL WORLD

1.1. What Is An Animal?

We can answer this question in two ways. One answer is relatively straight-forward and is provided in the following paragraph, and a first attempt at the really interesting answer occupies the rest of this book.

The animals are a monophyletic group of eukaryotes with a multicellular common ancestor that fed holozoically (i.e., on particulate food) and comprised cells that lacked a cell wall and included multiple morphologically and functionally different cell types. These traits set animals apart from the only other major group of ancestrally multicellular eukaryotes, the terrestrial plants, and other eukaryotes with multicellular representatives, e.g., the fungi and red algae, which have saprophytic or photosynthetic lifestyles and are composed of cells enclosed within a cell wall. Early in their diversification, two key innovations evolved in the animals: the gut, permitting exploitation of large food items; and the nervous system, laying the foundation for the complex behavior displayed by many animals. A further defining feature of the animals is that, as a group, they lack key metabolic capabilities common to various other organisms, including the capacity to photosynthesize, fix nitrogen, synthesize many amino acids contributing to protein, and produce various cofactors required for the function of enzymes central to metabolism. In other words, many animals are morphologically complex and some are clever, but all are metabolically impoverished.

This description of animals is not incorrect, but it is incomplete. It omits 1–10% of the biomass and half or more of the cells in the animal body. The missing cells are the microbial communities that live persistently with the animal. Most of these microorganisms are bacteria, but they also include unicellular eukaryotes; and large animals additionally bear multicellular eukaryotes, including mycelial fungi, mites, helminth worms, etc. Traditionally, these inhabitants of animals have been ignored unless they are injurious to the health of their animal host, and because many are difficult to culture, their ubiquity and diversity are grossly underestimated by routine culture-based microbiological methods.

The study of microorganisms associated with animals has been transformed by culture-independent methods to identify and study the function of microorganisms. The key technology has been high throughput DNA sequencing (also known as next generation sequencing), by which all the DNA, expressed genes, or specific genomic regions of interest in a sample can be sequenced simultaneously. It is now possible to determine the taxa in the microbial community and their functional traits, for example from a biopsy taken from the lung or intestine of a human patient, from a single soil microarthropod, or from the gills of a bivalve mollusk brought up from a deep sea hydrothermal vent. These ever-improving technologies have supported a decade or more of research on the microbiological natural history of animal bodies. It is now apparent that animals are the habitat for a previously unsuspected diversity and abundance of microbial residents, and this unfolding discovery has empowered experimental science, revealing that these microorganisms are critical to the health and well-being of their animal hosts.

So, what is an animal? It is a multiorganismal entity, comprising animal cells and microbial cells. The phenotype of an animal is not the product of animal genes, proteins, cells, tissues, and organs alone, but the product of the interactions between all of these animal functions with communities of microorganisms, whose composition and function vary with the age, physiological condition, and genotype of their animal host. Following from the growing appreciation of the significance of the microbiology of animals, many aspects of animal biology are being rewritten (McFall-Ngai et al., 2013). The biomedical sciences are increasingly recognizing the pervasive effects of resident microorganisms on human health. These effects extend beyond local impacts, for example of gut microorganisms on gut health and disease, to microbial effects on cardiovascular health, the integrity of the circadian rhythm, and psychiatric health. Many evolutionary biologists are realizing that the response of animals to selection is influenced by the impact of microbial partners on the trait under selection and the heritability of the microorganisms. Applied biologists are appreciating that microbes can shape the capacity of insects, such as mosquitoes, to vector disease agents (e.g., the malaria parasite, dengue virus) and their susceptibility to certain pest control agents. Furthermore, reliable predictions of the impacts of climate change on animal distributions will require consideration of the environmental requirements and dispersal capability of the microbial partners as well as the animal.

But, before we go any further, we need to address some issues of terminology. As with so many scientific disciplines, multiple terms are being used, sometimes interchangeably but often with different shades of meaning that can sow confusion and misunderstanding. Section 1.2 provides a guide to how some terms are used in this book, as well as why some terms are eschewed.

1.2. Terminology: Dismantling the Tower of Babel

Terms are important not only because they communicate agreed concepts within a discipline but also because they can encapsulate an entire conceptual framework. In this respect, the term "microbiome," coined to describe the catalog of microorganisms and their genes (Lederberg and McCray, 2001), is of central importance. The microbiome is a global, all-encompassing term for the microbiology of an animal, and is particularly useful where, as in shot-gun sequencing, individual genes cannot readily be assigned to particular microbial taxa. A related term is "microbiota," which refers specifically to the microbial taxa associated with an animal. The terms microbiome and microbiota are sometimes used interchangeably when referring to taxa. Alternatively, the microbiome can be used exclusively to refer to genes and genomes, with the microbiota as a taxonomic descriptor. It is usually obvious from the context how the term microbiome is being used. In this book, I will use both terms, with microbiome to describe inventories of genes, especially in relation to function, and microbiota when referring to the organisms.

The terms microbiome and microbiota have meaning for a science that is founded on molecular biology and genomics. A major driver of microbiome research over the last decade has been large consortium projects that have generated microbial sequences associated with humans, and their potential biomedical importance. However, microbiome research is also founded on many decades of pregenomic research on interactions between healthy animals and their resident microorganisms (Sapp, 1994). Although this research endeavor has been largely independent of biomedical science, the melding of the terminology of the pregenomic science of animal-microbial interactions with the terminology of -omic science has, to a large extent, been successful. But there have been some difficulties, and this is has caused some confusion and miscommunication.

A key pregenomic term is "symbiosis," which—as for microbiome—was invented to fulfil a scientific need. Symbiosis was coined in the 1870s to encapsulate the new discovery of multipartner organisms. It was initially used by Albert Franck to describe the composite nature of lichens, comprising a fungus and alga, and generalized in 1879 by Anton de Bary to describe the "living together of different species." Over the following century, most research on interactions between healthy animals and microorganisms focused on associations that were readily detected morphologically, including highly specialized interactions involving single microbial taxa housed in specific tissues or organs. These associations were categorized as symbioses (or sometimes endosymbioses), with the animal described as the host and microbial partner as the symbiont. The microbial symbionts in these "one-host-one-symbiont" or occasionally "one-host-two/three-symbionts" associations include the zooxanthellae (dinoflagellate algal cells) in corals, luminescent bacteria in the light organs of some fish and squid, and dense bacterial aggregations in specialized organs (bacteriomes) of certain insects. The text of Buchner (1965), together with two multiauthor volumes (Jennings, 1975; Nutman and Mosse, 1963), provide a superb overview of premolecular research on these associations.

Today, the terms microbiome and microbiota are generally used with reference to multitaxon microbial communities, and the microbial partner(s) tend to be called symbiont(s) where the interactions are limited to one or a few microbial taxa. There is a general presumption that, where the animal host is healthy, the terms microbiota and symbionts do not refer to pathogens. However, the impact of many microorganisms on the animal host can be context-dependent, varying with the developmental age, physiological condition, and genotype of the host, as well as environmental conditions. This was appreciated by Elie Metchnikoff who, in the early twentieth century, coined a further term "dysbiosis" as an antonym of symbiosis, to describe a microbial community that is deleterious to host health (Stecher et al., 2013). In the same vein, individual members of a microbial community that display context-dependent pathogenicity are often known as pathobionts (Hornef, 2015).

In the biomedical literature, the term "commensal" is widely used to describe individual taxa of the microbiota. This term poses some important problems. Strictly speaking, the term commensal refers to "eating at the same table," and has come to describe an organism that derives benefit from an association with no discernible effect on the fitness of its partner, akin to the sparrow feeding on the breadcrumbs dropped from a man's dining-table. The term commensal is not widely used in the symbiosis literature because it has all the standard difficulties of a negative definition: if only one used a more sensitive assay or studied the association under different conditions, perhaps benefit or harm would be detected, and the organism would, no longer, be a commensal. Microorganisms associated with the gut and skin of humans used to be called commensals, in the erroneous belief that they are of no significance. It is unfortunate that commensal has persisted into the era of microbiome research with the full knowledge that these microorganisms are crucial to the health of humans and other animals. Needless to say, verbal modifiers such as "beneficial commensal" further compound the terminological confusion. In the light of the complexities surrounding the term commensal, it is preferable to avoid this term. Alternative terms, such as microorganisms, microbial communities, etc. are sufficient; where it is important to emphasize that pathogens are specifically excluded from consideration, the term "nonpathogenic microorganisms" can be used to avoid any ambiguity.

There is one further set of terms that needs to be addressed: holobiont and hologenome. Lynn Margulis coined new terms for the partners in a symbiosis as "bionts" and the association as a "holobiont" (Margulis, 1991) to emphasize the evolutionary persistence of the association and how selection may operate at the level of the association (or holobiont). This terminology has been brought into the -omic era with a further new term, the hologenome, which refers explicitly to the host genome plus microbiome as the unit of selection (Rosenberg and Zilber-Rosenberg, 2016). The concept of the hologenome is relevant to a very restricted set of associations. In particular, it does not apply to the complex microbial communities that are the focus of much microbiome research, where individual microbial taxa have different selective interests from each other and variable selective overlap with the host. Strong overlap of selective interest between the partners is predicted in some associations involving individual microbial partners that are vertically transmitted (and so have a selective interest in the fitness of the host offspring), but the residual selective conflict limits the applicability of the hologenome concept. Even the most ancient of symbioses, between the eukaryotic cell and the mitochondrion, is subject to genomic conflict (Perlman et al., 2015) and so cannot be classified as a pure hologenome. For these reasons, which are elaborated further by Douglas and Werren (2016) and Moran and Sloan (2015), the hologenome concept is not developed in this book.

1.3. The Microbiology of Animals

Now that the main terms for the discipline are defined, we can consider why animals support microbial communities. The functional explanations are twofold: microorganisms provide metabolic capabilities that are lacking in animals; and microorganisms modulate the signaling networks that regulate animal functions required for sustained animal health and vigor (figure 1.1).

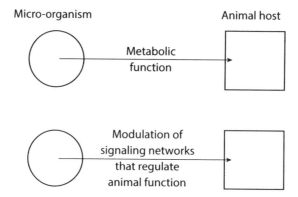

FIG. 1.1. How microorganisms interact with the animal host.

The key metabolic capabilities gained by animals from microbial partners are listed in Table 1.1A. Some of these capabilities were absent from the lineage giving rise to animals, e.g., photosynthesis, nitrogen fixation, and essential amino acid and B vitamin synthesis, and the dietary requirement of various animals for carbon, nitrogen, or specific nutrients has been spared by symbiotic microorganisms. For example, photosynthetic associations have evolved repeatedly in basal animals; some termites possess nitrogen-fixing bacteria in their guts, enabling them to thrive on wood of exceptionally low nitrogen content; and all animals that feed through the life cycle on vertebrate blood, e.g., bedbugs, tsetse flies, leeches, and ticks, are widely believed to derive supplementary B vitamins from microbial partners (Douglas, 2015; Venn et al., 2008).

Symbiotic microorganisms also complement metabolic deficiencies that have evolved in specific animal groups. For example, arthropods cannot synthesize sterols, which are essential constituents of membranes and various hormones, and some insects living on sterol-poor diets derive most of their sterol requirement from yeast symbionts (Douglas, 2015); and, because vertebrates cannot degrade cellulose, most herbivorous vertebrates depend on cellulolytic microorganisms in their guts. Some metabolic traits are mediated either intrinsically (i.e., by the products of animal genes) or via microbial partners, varying among animal taxa. Complex patterns of intrinsic and microbial origins of metabolic traits are evident for cellulose degradation (Calderon-Cortes et al., 2012). Similarly, animal luminescence can be intrinsic or microbial. An inventory of light production in marine fish identified 8 independent origins of intrinsic luminescence and 17 origins of bacterial luminescence (Davis et al., 2016). This includes the deep-sea angler fish that maintain two light organs, one with intrinsic luminescence and the other housing symbiotic bacteria. Secondary metabolism, including the synthesis and degradation of toxins, is also mixed in origin (Table 1.1A)

TABLE 1.1. Services Provided by Microbial and Animal Partners

(A) Microbial services	Examples
CO_2 fixation	Dinoflagellate algae Symbiodinium in shallow-water scleractinian corals
—Photosynthesis	Bacteria in annelid tube worms and various
—Chemosynthesis	bivalve mollusks at hydrothermal vents and methane seeps
Nitrogen fixation	Various bacteria in some termites
Essential amino acid synthesis	Various bacteria in plant sap–feeding insects
B vitamin synthesis	Various bacteria in insects feeding through the life cycle on blood
Degradation of complex polysaccharides	Bacteria in anoxic gut regions of vertebrate herbivores
Luminescence	*Vibrio* or *Photobacterium* in light organs of some fish and squid
Toxin (e.g., polyketide, antibiotic) synthesis	Various bacteria in benthic marine animals, e.g., sponges, bryozoans, and insects (e.g. *Paederus* rove beetles, attine ants)
Toxin degradation	Various gut bacteria in animals feeding on toxic plant tissues, e.g., detoxification of plant phenolics ingested by desert woodrats, caffeine ingested by the coffee berry borer beetle

(B) Animal services	Examples
Nutrient supply: selective feeding harvests and concentrates nutrient-rich substrates that are utilized by microorganisms	Many gut microorganisms
Protection from adverse abiotic conditions, e.g., high oxygen tensions, desiccation	Obligately anaerobic microorganisms in anoxic fermentation chambers of animal guts
Provide enemy-free space by sequestering microorganisms and by immune system function	Inferred for many intracellular microorganisms
Dispersal, often in nutrient-rich substrate, e.g., mucus, fecal material	Inferred for carriage of some microorganisms through the animal gut

although, as the microbiology of more animals is investigated (e.g., Ceja-Navarro et al., 2015; Florez et al., 2015; Kohl and Dearing, 2016), many more instances of microbial-mediated secondary compound metabolism may be revealed. The evolutionary and ecological factors that determine why some traits can be either intrinsic or microbial origin have not been investigated systematically.

The many examples of animals that gain access to metabolic capabilities through associations with microorganisms can be explained in terms of an inherited predisposition of animals to associate with microorganisms, especially bacteria. This evolutionary predisposition appears to be common to all eukaryotes. The common ancestor of all modern eukaryotes bore an intracellular *Rickettsia*-like bacterium that evolved into the mitochondrion; unicellular eukaryotes (protists), where investigated, are routinely colonized by bacteria or other microorganisms; the plant microbiome is a very active area of current research (Bai et al., 2015; Lundberg et al., 2012); and discovery of additional fungal or bacterial partners in lichenized and mycorrhizal fungi illustrates how fungi also have a propensity to form associations (Bonfante and Anca, 2009; Spribille et al., 2016). This microbio-philia of animals appears to be matched by advantages to the microorganisms of colonizing the animal, including access to nutrients, protection from natural enemies or harsh abiotic conditions, as well as dispersal (Table 1.1B).

These considerations bring us to the second reason why animals are associated with microorganisms. Because animals originated and diversified in the context of a long evolutionary history of relationships with microorganisms, the key physiological systems of animals, together with the signaling networks that regulate these systems, all evolved in the context of preexisting and ongoing interactions with microorganisms. In other words, the microbiome is expected to play a role, directly or indirectly, in the development and function of the animal nervous system, immune system, endocrinal system, gut physiology, respiratory physiology, and so on. This reasoning takes us beyond the predictions that microorganisms confer various metabolic services, as outlined in Table 1.1A, to the additional prediction that the presence and activities of microorganisms influence many, possibly all, aspects of animal function (figure 1.1B).

Just as the role of microorganisms as a source of metabolic capabilities (figure 1.1A) has been appreciated for many years, the predicted integration of microorganisms into all aspects of animal function also has a strong historical basis. It was Louis Pasteur, the architect of the germ theory of disease, who first addressed the significance of microorganisms for animals, arguing that microorganisms are essential for animal life (Pasteur, 1885). We now know that this is not true, and that some animal species can be maintained under microbiologically sterile conditions; they are known as "axenic" or "germ-free" animals. The viability of germ-free animals was first demonstrated for *Calliphora* blow flies (Wollman, 1911), and the subsequent elucidation of the nutritional requirements of *Drosophila* was founded on methods devised to eliminate all microorganisms from the fly cultures (Sang, 1956). Today, the production and distribution of germ-free laboratory mice (Smith et al., 2007) is a commercial enterprise. These germ-free mice display many symptoms of ill-health, including stunted growth, depressed fertility, and reduced metabolic rate, commonly accompanied by specific abnormalities of various organs. These multiple deficiencies and abnormalities are the basis for the fundamental role of microorganisms in animal health and well-being. In other words, we can only understand animal function by integrating the microbiology of animals into our explanations of animal biology.

1.4. Scope of This Book

The realization that every animal is colonized by microorganisms that can shape its health and well-being is transforming our understanding of animal biology. The purpose of this book is to provide some initial explanations and hypotheses of the underlying animal-microbial interactions. For this, we need multiple disciplinary perspectives.

We start with evolutionary history in chapter 2. The propensity of animals to associate with microorganisms has ancient roots, derived from both the predisposition of all eukaryotes to participate in associations and, likewise, the tendency of many bacteria to interact with different organisms, often to mutual benefit. Chapter 2 outlines the patterns of these interactions, especially in taxa related to animals and basal animal groups. Interactions are mediated by chemical exchange, enhancing access to energy and nutrients and providing chemical information that enables the interacting organisms to anticipate and respond adaptively to environmental conditions. Many of these core interactions were firmly established in the ancestor of animals. The multicellular condition of animals, sophisticated immunological function of even basal animals, and key animal innovations, including the polarized epithelium and the gut, play important roles in shaping the pattern of animal-microbial interactions.

Although all animals are associated with microorganisms, we know more about the microbiome of humans than any other animal. Chapter 3 addresses current understanding of the role of the microbiome in human health. Studies of the microbiology of humans combined with experimental analyses of model animals are revealing complex problems—and some solutions. The complexity lies in the great diversity of microorganisms within each individual human, as well as considerable among-individual variation; and the importance of the microbiome is reinforced by the increasing evidence for microbial involvement in some diseases, especially metabolic and immunological dysfunctions. Western lifestyles, including diet and antibiotic treatment, have been argued to contribute to the incidence of microbiome-associated diseases, with opportunities for microbiological restoration by microbial therapies.

Our understanding of interactions between animals and the microbiome is most developed in relation to the immune system, and this is the focus of chapter 4. It is now apparent that animal immune system is a key regulator of the abundance and composition of the microbiota, and that immunological function is strongly regulated by the composition and activities of the microbiome. The immune system cannot be understood fully except in the context of the microbiology of the animal. Furthermore, this highly interactive system is overlain by microbial-mediated protective functions, essentially comprising a second immune system.

Chapter 5 investigates the role of the microbiome in shaping animal, including human, behavior. It has long been known that pathogens can drive animal behavior, and there is now increasing evidence that resident microorganisms can have similar, although often more subtle, effects. Research has focused primarily on three aspects of animal behavior: feeding behavior, chemical communication among animals, especially in relation to social interactions, and the mental well-being of mammals, including humans. As chapter 5 makes clear, this topic has attracted tremendous levels of interest, but fewer definitive data.

The impacts of animal-associated microorganisms on host health and their interactions with the immune system and nervous system of animals (chapters 3–5) have one overriding theme in common: that these interactions are complex, with multiple interacting variables. This complexity can often appear to defy comprehension. Chapter 6 discusses the ecological approaches that have the potential to solve many of these problems of complexity. Treating the animal as an ecosystem, we can ask multiple questions: what are the ecological processes that shape the composition and diversity of microbial communities, and how do these properties of the microbial communities influence overall function of the ecosystem? Research on complex microbiomes, especially in the animal gut, as well as one-host-one-symbiont systems are revealing the role of interactions among microorganisms and interactions between the microorganisms and host in shaping the diversity of the microbiome. Furthermore, the response of individual taxa and interactions can influence the stability of communities to external perturbations, ranging from the bleaching susceptibility of shallow-water corals to the gut microbiota composition of humans administered with antibiotics.

In chapter 7, the evolutionary consequences of animal-microbial associations are considered. There is a general expectation that the fitness of both animal and microbial partners is enhanced by these associations largely through the reciprocal exchange of services. Nevertheless, hosts can exploit their microbial partners, and there are indications that animals can be addicted to their microbial partners. At a broader scale, this chapter investigates how these associations affect the rate and pattern of evolutionary diversification of the microbial and animal partners. In addition to evidence for coevolutionary interactions and facilitation of horizontal gene transfer, various studies point to a direct role of microbiota in interrupting gene flow and speciation by both prezygotic and postzygotic processes.

Finally, chapter 8 addresses the implications of the microbiology of animals and some key priorities for future research. It is now abundantly clear that the microbiome has pervasive effects on the physiological and developmental systems of animals and the resultant animal phenotype. One of the big biological questions in the life sciences today concerns how the phenotype of an animal maps onto its genotype and the underlying physiological and developmental mechanisms. The answers to this question will require the integration of the microbiome with the traditional animal-only explanations of animal function. As this book illustrates, the technologies and concepts to achieve this intellectual transformation of animal biology are largely in place. Why is this integration of disciplines needed? Beyond the fundamental priority to understand and explain, the microbiome offers important, but currently untapped, routes to promote human health and to mitigate and manage some of the damaging effects of human activities on our environment.

2

The Ancient Roots of the Animal Microbiomes

2.1. Introduction

Our ancestors were multiorganismal before they were multicellular. In other words, the propensity of animals to associate with microorganisms is an inherited trait shared with their unicellular relatives, and did not evolve in concert with, or after, the origin of the multicellularity. The implications of the ancient evolutionary roots of animal-microbial associations are profound. Key animal innovations, ranging from the gut and nervous system to the adaptive immune system and complex behavior, arose and diversified in the context of preexisting networks of interactions with microorganisms. Consequently, any explanation of the function of the tissues and organs of animals requires consideration of microbial dimension, as well as the animal genetic repertoire and functional organization that are used in traditional explanations.

The purpose of this chapter is to outline the evidence for the ancient roots of the microbiology of animals, and the likely patterns of interactions that animals have inherited from their microbial ancestors. It is becoming increasingly evident that the capacity to communicate with other organisms is a universal feature of living organisms, and section 2.2 of this chapter addresses signal exchange in bacteria. We then turn to the eukaryotes, addressing the evidence that the common ancestor of all modern eukaryotes was not a single cell but a symbiosis between two microbial cells (section 2.3); the ubiquity and diversity of associations in unicellular eukaryotes, including the closest relatives of extant animals, the choanoflagellates (section 2.4); and, finally, the associations with microorganisms in the basal groups of animals, setting the scene for the diversification of complex tissues and organs in various animal groups, including mammals and humans (section 2.5).

2.2. The Social Life of Bacteria

2.2.1. CHEMICAL COMMUNICATION AMONG BACTERIA

This planet has borne living organisms for billions of years, likely stretching back at least 3.6 billion years. Prior to the first fossil evidence for eukaryotes within the last 1–2 billion years, the biosphere was dominated by organisms akin to modern bacteria (figure 2.1A), and many microbiologists argue that bacteria are the dominant life forms to the present day. The term "bacteria" is a grade of organization that comprises two major domains, the Eubacteria (sometimes, confusingly, referred to as the Bacteria) and the Archaea (figure 2.1B). In comparison to eukaryotes, the morphological diversity of bacteria is relatively limited. They are generally small (cell size rarely exceeds 10 µm) and occur as single cells or short chains or filaments. Metabolically, however, bacteria are tremendously varied, and much of the metabolic innovation is very ancient, preceding the first eukaryotes in the fossil record. The metabolic capabilities of bacteria involve multiple routes to harness chemical and light energy for carbon dioxide fixation (chemosynthesis and photosynthesis, respectively), to obtain energy by the oxidation of organic compounds (e.g., sugars, methane) and inorganic compounds (e.g., sulfide, Fe[II]), and to fix atmospheric nitrogen into ammonia. Oxygen-evolving photosynthesis, which evolved ca. 3 billion years ago, probably in the ancestors of modern cyanobacteria, was a run-away success, resulting in global pollution of the atmosphere with molecular oxygen (figure 2.1A) and the selection pressure for the evolution of aerobic respiration, with oxygen as the terminal electron acceptor.

The principal method that enabled microbiologists in the twentieth century to discover the great metabolic diversity of the bacteria was analysis of bacteria in pure culture, where genetically uniform bacterial cells are grown planktonically in nutrient-rich liquid with minimal cell-cell contact. The methodology has been enormously powerful, but it also discouraged any serious consideration of among-bacterial interactions. This important gap in our understanding has, however, been redressed, especially in the last decade or so. It is now very evident that bacteria have a rich and varied social life, and that this among-bacterial communication is mediated largely by transfer of organic molecules. The chemical systems can be organized conveniently as involving either primary metabolites required for survival

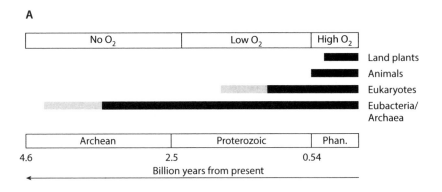

A

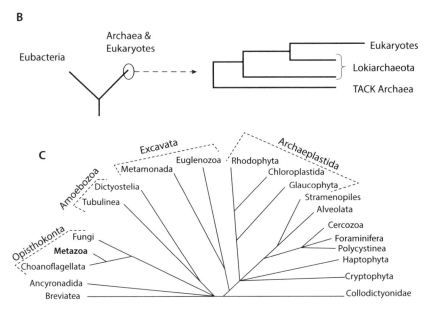

FIG. 2.1. The diversity of living organisms. A. Evolutionary history of life (solid bars, confident interpretation of geological record; gray bars, fragmentary or controversial evidence) Phan. = Phanerozoic, i.e., from Cambrian era to present. Eukaryotes = derived from the last common ancestor of modern eukaryotes (which would have possessed mitochondria). (Modified from Fig. 3 of Knoll [2014].) B. The primary division of living organisms into two domains, the Eubacteria and Archaea, with the eukaryotes likely derived from within the archaeal phylum Lokiarchaeota, allied to the superphylum TACK (Thaumarchaeota, Aigarchaeota, Crenarchaeota, and Korarchaeota). C. The diversity of eukaryotes, with the Metazoa (animals) in bold. The phylogenetic relationships in the eukaryotes, especially among many unicellular forms (informally known as protists) are not firmly established, and are likely to undergo further revisions. *The Chloroplastida includes the terrestrial plants.

and growth or secondary metabolites, which are compounds with ecological and social functions. Although, like many classification schemes, the distinction between primary and secondary metabolites can occasionally be blurred, it provides a useful framework for considering chemical communication among bacteria.

2.2.2. SHARING PRIMARY METABOLITES: SYNTROPHY

Primary metabolites are shared widely among bacterial cells. In many instances, this is a consequence of the small size and large surface area/volume relationship of bacterial cells, resulting in incomplete control over the release of intracellular metabolites. In some circumstances, the cross-feeding of metabolites, also known as syntrophy, between bacteria with different metabolic capabilities can be mutually beneficial. Syntrophy has traditionally been viewed as important for bacteria in energy-limited, anoxic environments, where various bacteria with complementary metabolic capabilities form intimate associations (Morris et al., 2013). For example, fermentative bacteria, which produce hydrogen as they oxidize organic compounds (e.g., ethanol to acetate), are commonly associated with methanogens that consume the hydrogen as substrate for methane production, acting as an electron sink favoring sustained fermentation by its partner (figure 2.2A). The burgeoning information on microbial communities in diverse habitats obtained from genomic and metagenomic analyses is revealing that syntrophic interactions may be more widespread than traditionally appreciated, and may occur in nutrient-rich environments where there is temporal variability in nutrient availability. For example, Zelezniak et al. (2015) used genome-scale data to compute the metabolic interactions in bacterial communities from >800 habitats, revealing a high predicted incidence of cooperative metabolite exchange among two, three, or four taxa (larger bacterial consortia were not investigated). These interactions commonly involved bacteria of different phyla, especially Actinobacteria, Proteobacteria, and Firmicutes, with carbohydrates and amino acids particularly likely to be exchanged. The authors of this study raise the possibility that, although the taxa may vary among the different habitats, there may be "recurring modules" of cooperative groups of bacteria that contribute to the architecture of microbial communities in diverse habitats.

Some syntrophies are likely ancient and highly coevolved. One example is a two-member bacterial association in symbiosis with a group of insects known as sharpshooters. As figure 2.2B illustrates, the synthesis of the amino acid methionine by one bacterium, *Baumannia*, is dependent on the provisioning of the precursor homoserine from the second bacterium, *Sulcia*. However, studies on various synthetic associations suggest that bacteria can engage in mutually advantageous metabolic interactions without a long coevolutionary history. One study (Pande et al., 2015) used *E. coli* and *Acinetobacter baylyi* genetically engineered to overproduce one amino acid but to require a second amino acid (tryptophan and histidine). The two bacteria could be cocultured on medium lacking both amino acids, when the *E. coli* cells overproduced histidine and required tryptophan and the *A. baylyi* cells required histidine and overproduced tryptophan, or vice versa. The bacterial cells used in these experiments bore plasmids encoding genes for the synthesis of different fluorescent proteins, and the cross-feeding of amino acids was accompanied by the transfer of fluorescent proteins between the different bacteria, suggesting that the amino acids may have been transferred by cytoplasmic exchange between the bacteria. Scanning electron microscopy revealed tubular structures ("nanotubes") between the bacterial cells, although further research using live imaging is needed validate these structures.

2.2.3. INFO-CHEMICALS IN THE BACTERIAL WORLD

In addition to the hundreds of primary metabolites, bacteria synthesize a wide diversity of secondary compounds (Davies and Ryan, 2012). Some secondary compounds are toxic to other organisms (they kill or debilitate the recipient

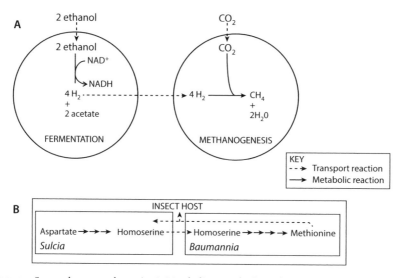

FIG. 2.2. Syntrophy among bacteria. A. Metabolic cross-feeding of hydrogen from fermentative bacteria to methanogens: Reactions yielding hydrogen enable the recovery of NADH in fermentative bacteria and the removal of hydrogen by hydrogen-scavenging microorganisms (methanogen, shown, acetogens, or sulfate-reducing bacteria), which promotes the overall efficiency of the fermentation reaction. B. Metabolite cross-feeding between two symbiotic bacteria, *Sulcia* and *Baumannia*, in the glassy-winged sharpshooter *Homalodisca vitripennis* mediates the net synthesis of the amino acid methionine, which is required for protein synthesis by both bacteria and the insect host. (Data from McCutcheon and Moran [2007].)

cell) but they can also function as info-chemicals, i.e., they induce phenotypic changes that are advantageous to the recipient, mediated by changes in gene expression patterns, conformation of specific proteins, or membrane permeability and the ionic balance of the cell. From a functional perspective, info-chemicals can be classified as either cues or signals (figure 2.3A; for a general consideration of the distinction between signals and cues, see Maynard Smith and Harper [2003] and Padje et al. [2016]). Info-chemicals that are produced for functions that are independent of their effect on the recipient are known as cues. From the perspective of the recipient, a cue is a reliable indicator of environmental conditions, e.g., the smell of a predator. A signal, by contrast, mediates a coevolved interaction between the signaler and recipient: the signal has evolved because of its effect on other organisms and is effective only because the response of the recipient has also evolved.

The best studied class of bacterial info-chemicals is the quorum sensing (QS) molecules, by which bacterial cells sense density. For a population of bacteria of single genotype, the capacity to sense density is important because some bacterial functions are adaptive only at high density. For example, light production by luminescent bacteria is costly (it consumes ATP) and, because high densities of luminescent bacterial cells are required for the production of detectable amounts of light, it is adaptive for expression of luminescence genes to be suppressed in dilute suspensions of bacteria. A more widely distributed trait of bacteria is the capacity to form biofilms, i.e., aggregations of cells within an extracellular matrix, known as EPS (extracellular polymeric substances) comprising a complex, hydrated mix of polysaccharides and proteins. As with light, EPS is costly to produce and only advantageous at high bacterial densities. Interestingly, many of these traits displayed at high bacterial densities are "public goods," meaning that they are available for use by all cells in the community and are not depleted. In other words, the production of light or EPS is a social trait of bacteria, and it is in the interests of each luminescent or EPS-secreting cell that all the other cells are doing likewise. QS coordinates this cooperative behavior. Specifically, every bacterial cell releases a signaling molecule which accumulates at a rate proportional to the bacterial density. At a certain threshold density of bacterial cells (analogous to the quorum of people required for a committee to make binding decisions), the signaling molecule induces the expression of genes that mediate social functions. The QS molecules vary among bacterial species, and include acyl homoserine lactones in many Gram-negative bacteria, and modified peptides or amino acids in many Gram-positive bacteria.

QS-mediated communication among bacteria of the same or related genotypes is an example of signaling because all the cells derive benefit (all the cells are both producers and recipients). The cells are adapted to display a change in phenotype on receipt of the QS molecules at concentrations greater

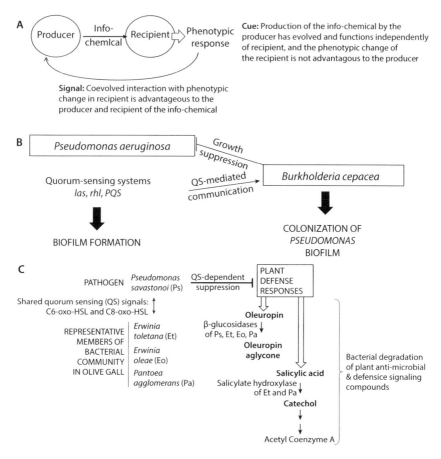

FIG. 2.3. Chemical communication in bacteria. A. Info-chemicals include cues and signals. B. QS-mediated communication between *Pseudomonas aeruginosa* and *Burkholderia cepacia*. The QS systems of *P. aeruginosa* function interactively to mediate density-dependent biofilm formation, and QS molecules are used as a cue by *B. cepacia* for colonization, forming two-species biofilms in co-culture *in vitro*, and suppressing *P. aeruginosa* populations. C. Interactions in multibacterial community occupying galls on olive trees. The severity of the olive knot disease caused by the pathogen *P. savastanoi* pv. savastanoi is enhanced by co-occurring bacteria, through shared quorum sensing signaling (HSL: homoserine lactone) and shared degradation of plant defensive compounds, including the production of β-glucosidases that degrade the plant phenolic gluco-side, oleuropein, and salicylate hydroxylase, which inactivates salicylic acid, a key antimicrobial signaling molecule of plants. (Fig. 2.3C modified from Fig. 1 of Buonaurio et al. [2015].)

than a certain threshold, and this change in phenotype is advantageous to the cells producing the QS molecules. QS molecules also function, usually as a cue, in interspecific communication. For example, QS plays a central role in communication between *Pseudomonas aeruginosa* and *Burkholderia cepacia* (figure 2.3B), both of which form chronic infections in the lungs of cystic fibrosis patients (Riedel et al., 2001). *P. aeruginosa* uses three linked

QS systems (*las, rhl,* and *PQS*) for density-dependent control of the expression of ca. 300 genes that mediate biofilm formation and the production of various virulence factors (Parsek and Greenberg, 2005). In cocultures, the QS molecules of *P. aeruginosa* induce *B. cepacia* to participate in the biofilm and produce virulence factors. *B. cepacia* utilizes the QS molecules of *P. aeruginosa* as a cue (not a signal) because its production by *P. aeruginosa* is independent of the presence of *B. cepacia* and *P. aeruginosa* derive no discernible benefit from co-colonization by *B. cepacia*. Indeed, there is evidence that *B. cepacia* inhibits the growth and reduces the abundance of *P. aeruginosa* in the lungs of cystic fibrosis patients, as well as exacerbating the disease symptoms of the patient (Schwab et al., 2014).

Among-species communication via QS has also been demonstrated in relation to another pathogenic *Pseudomonas* species: *P. savastanoi* pathovar savastanoi, which forms galls ("olive knots") on the shoots of olive trees, reducing olive production and, when a gall girdles a branch, causing dieback. The cavity of the gall is generated by plant cell wall–degrading enzymes of *P. savastanoi*, and is occupied by the bacterial cells. However, the galls are larger and disease severity greater where the *P. savastanoi* cells are accompanied by other bacteria that are harmless to the plant in the absence of *P. savastanoi* and that, collectively, can account for up to 50% of the bacterial cells in the gall. As a consequence of among-bacterial communication via QS, these latter bacteria promote the capacity of *P. savastanoi to* suppress plant defenses (figure 2.3C). In particular, a frequent coinhabitant of the galls, *Erwinia toletana* produces the same two QS molecules as *P. savastanoi*, and therefore *P. savastanoi* cells are virulent to the plant host at lower densities when co-occurring with *E. toletana* (Buonaurio et al., 2015). In contrast to the interactions between *P. aeruginosa* and *B. cepacia* (figure 2.3B), the interspecies QS communication between *P. savastanoi* and *E. toletana* is likely signaling because, as in single-species QS interactions, both producer and recipient of the QS molecules derive benefit from the interaction. The diverse bacteria also display the cooperative production of degradative enzymes that break down antibacterial products of the host (figure 2.3C), but it is currently unknown whether these traits are under joint QS control.

As these examples of both primary and secondary metabolites illustrate, many bacteria are generally predisposed to interact. Relationships range from cells of a single genotype to highly divergent taxa, and from flexible, context-dependent interactions to intimate coevolved associations. The participating bacteria commonly differ in their metabolic capabilities and, while many interactions are mediated by the exchange of diffusible compounds, others involve the transfer of membrane vesicles or cytoplasmic contents. It is in this milieu of complex among-bacterial interactions that the eukaryotes evolved.

2.3. The Multiorganismal Origins of Eukaryotes

2.3.1. THE ANCESTORS OF THE EUKARYOTIC GENOME

The eukaryotes, i.e., organisms whose cells have a double membrane–bound nucleus, comprise a very substantial adaptive radiation, mostly of taxa that are unicellular for all or most of their life cycle (figure 2.1C). Although the root of the eukaryotic tree is enigmatic, analysis of the sequences of the slowly evolving rRNA genes indicate clearly that the eukaryotes evolved from the Archaea (figure 2.1B), probably from within (or allied to) the candidate phylum Lokiarchaeota, which is currently known only from sequence data of uncultured cells sampled from deep sea sediments (Spang et al., 2015). The sequence of many eukaryotic genes, especially those involved in DNA replication, transcription, and repair, and in protein translation, are allied with archaeal genes, but other genes, including many metabolism-related genes, appear to be related to eubacterial genes (Williams et al., 2013). The variation in phylogenetic signal from different genes is likely a consequence of horizontal gene transfer, which is widespread both between Eubacteria and Archaea and increasingly recognized in eukaryotes. In other words, the ancestral genome of modern eukaryotes has multiorganismal origins.

Importantly for our understanding of animal origins, the multiorganismal nature of eukaryotes extends beyond gene content of the genome to cellular organization. This is because all modern eukaryotic cells have evolved from a symbiosis with an intracellular *Rickettsia*-like bacterium, as we consider next.

2.3.2. THE SYMBIOTIC ORIGIN OF MODERN EUKARYOTES

All extant eukaryotes have evolved from an ancestor with bacterial symbionts. The basis for this statement is the overwhelming evidence that all modern eukaryotes either possess mitochondria or evolved from mitochondriate ancestors, and that the mitochondria arose from a single intracellular α-proteobacterial symbiont broadly allied to modern *Rickettsia*. How the origin of this association relates to the origin of eukaryotes is disputed (Martin et al., 2015). Perhaps the symbiosis with the ancestor of mitochondria co-occurred with, and may even have been required for, the evolution of the eukaryotic cell. Alternatively, the eukaryotic condition may have evolved first, and the eukaryotes that did not acquire mitochondria went extinct—or have yet to be identified. For our purpose, however, the important point is that the animals comprise one lineage within an adaptive radiation of eukaryotes whose common ancestor bore an intracellular bacterium. The modern eukaryotic cell has been designed through natural selection to interact with bacterial cells.

A widespread trait of animal-microbial associations is displayed by the relationship between mitochondria and the eukaryotic cell: the host (the animal or eukaryotic cell) preferentially associates with microbial partners that possess a capability that is both absent from, and advantageous to, the host (see chapter 1, section 1.3 and Table 1.1A). The capacity to produce energy at high rates via the membrane-bound electron transport chain and oxidative phosphorylation evolved ca. 2 billion years ago, in response to the increasing availability of free oxygen in the environment (figure 2.1A), but was absent from the archaeal lineage giving rise to the eukaryotes. The association with the *Rickettsia*-like bacterium gave the ancestral eukaryote access to the products of oxidative phosphorylation, enabling it to escape from the severe restrictions on energy production imposed by substrate-level phosphorylation.

At first sight, the selective advantage of oxidative phosphorylation to the eukaryotic host might be envisaged to relate primarily to competitive interactions, especially with free-living bacteria possessing this metabolic capability. There is, however, a further advantage that may have been of far greater evolutionary significance: the association relieved two important constraints faced by bacteria with a cell membrane–bound electron transport chain. The first is the constraint on bacterial cell size, a consequence of the capacity to produce energy being determined by the surface area of the cell, which increases by 2/3 for every doubling in cell volume. Unlike a bacterial cell, a mitochondriate eukaryotic cell can combine large total cell volume with high rates of energy production mediated by the large surface area of ATP-generating membrane provided by multiple internal mitochondria (figure 2.4A). The second relates to the function of the cell membrane, which plays a crucial role in defining how the cell interacts with the environment, by virtue of its location at the interface between the cell contents and external environment. Variation in bacterial cell membrane function is limited by the tight physico-chemical requirements of membrane composition and charge required for electron transport and ATP synthesis. In contrast, the eukaryotic cell membrane can be mechanically dynamic, endocytically active, and variable in composition and function, and consequently can interact with the external environment in diverse ways. The resultant division of labor between the membranes of the eukaryotic cell and its intracellular bacteria is, arguably, the basis of the adaptive radiation of eukaryotes, with ecological opportunities unavailable to the archaeal and eubacterial ancestors of the eukaryotic cell and its mitochondria.

In many ways, the mitochondrion is a special case among the many associations between eukaryotes and microorganisms because it is an organelle. Through a combination of relaxed selection and genomic decay, associated with large-scale transfer of genes to the eukaryotic nucleus, the mitochondrial genome in most eukaryotes is at least an order of magnitude smaller than the 1–1.5 Mb genomes of modern *Rickettsias* (exceptionally, the mitochondrial genomes of some land plants are ≥ 1 Mb through secondary expansion of noncoding regions) (figure 2.4B). For example, the human mitochondrion (16.6 kb) has just 13 protein-coding genes, all contributing to the electron transport chain, whose synthesis requires 24 rRNA and tRNA mitochondrial genes; and its function is underpinned by the import of ca. 1,500 proteins coded in the nucleus and synthesized in the cytoplasm of the eukaryotic host cell (Schmidt et al., 2010). Even more extreme reductive evolution of the mitochondrion is evident in various anaerobic protists, where the membranes of the mitochondrion are retained, but the genome is entirely lost. The resultant organelle functions to synthesize ATP via pyruvate ferredoxin oxidoreductase and hydrogenase, and is termed a hydrogenosome (Embley et al., 2003).

The evolutionary transition from bacterium to organelle is not, however, unique to the mitochondrion, but the incidence of bacterium-to-organelle transitions is uncertain, and depends on the criteria used to define a bacterial-derived organelle. Extreme genome reduction is universally accepted as a defining feature of a bacterial-derived organelle. Additional criteria include subsidy of organelle function by proteins imported from the cytoplasm of the eukaryotic cell and that the genes coding some of these proteins have

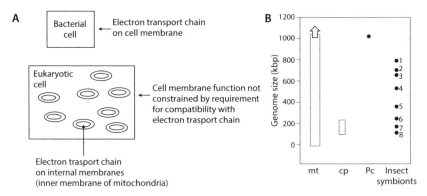

FIG. 2.4. Mitochondria and other candidate bacterial-derived organelles. A. The localization of the electron transport chain to internal membranes (mitochondria) facilitates large size and complex cell membrane function of eukaryotic cells. B. Genome sizes of: mitochondria (mt) from 6 kbp in *Plasmodium* to >1 Mbp in some land plants (e.g., *Silene conica*, 11.3 Mbp); plastids (cp) 110–200 kbp, the chromatophore of the amoeba *Paulinella* (Pc) 1.02 Mbp; and various intracellular symbionts of insects: 1 *Blochmannia* PENN 791 kbp, 2 *Wigglesworthia* GB 703 kbp, 3 *Buchnera* APS 656 kbp, 4 *Moranella* PCIT 538 kbp, 5 *Portiera* BTB 358 kbp, 6 *Sulcia* GWSS 245 kbp, 7 *Tremblaya* PAVE 172 kbp, 8 *Nasuia* NLF 112 kbp. The endosymbionts with very small genomes <2–3 kbp are sometimes considered as bacterial-derived organelles, broadly equivalent to mt and cp, and symbionts with larger genomes are generally regarded as bacteria (genome size data collated from NCBI).

been transferred horizontally from the bacterial ancestor of the organelle to the nucleus of the host cell. By all three criteria, two different cyanobacterial ancestors have given rise to plastids, one in almost all modern photosynthetic eukaryotes, and the other in amoebae of the genus *Paulinella* (McFadden, 2014; Nowack et al., 2016). Various intracellular symbionts in insects have very small genomes (figure 2.4B, see also chapter 7, section 7.3.2 where the evolutionary processes underlying extreme genome reduction are discussed) and they are sometimes referred to as organelles, even though their function does not appear generally to depend on import of host-derived proteins. As the microbiology of other eukaryotes, including animals, is studied further, other instances of extreme genome reduction and organelle formation may be revealed.

2.4. The Ubiquity of Microbial Associations in Eukaryotes

2.4.1. COMMUNICATION BETWEEN PROTISTS AND BACTERIA

Not only does the relationship between the first eukaryotes and mitochondria demonstrate the ancient propensity of eukaryotes to form associations with bacteria, but it has also facilitated the capacity of eukaryotic cells to interact with other organisms. Because of their large size (most eukaryotic cells are 10–100 μm diameter, compared to 0.1–5 μm for most bacteria) and endocytically-active cell membrane, eukaryotic cells provide multiple habitats for bacteria. Bacteria can form substantial, often multispecies colonies on the surface of eukaryotic cells, and also colonize the cytoplasm and occasionally eukaryotic organelles, including the nucleus, endoplasmic reticulum, and mitochondria.

From these considerations, we should expect associations with bacteria to be very widespread, perhaps ubiquitous, among unicellular eukaryotes (protists, see figure 2.1C). Consistent with this prediction, microscopical and molecular analyses generally concur that apparently healthy protists are routinely colonized by bacteria (Nowack and Melkonian, 2010; Schulz and Horn, 2015). Although the incidence of associations has not been investigated systematically across the phylogenetic diversity of protists, several systems have been studied in detail. In particular, various protists gain nutritional or defensive benefits from interacting with bacteria. Marine ciliates of the genus *Euplotidium* bear on their dorsal surface a dense colony of bacteria that are allied to Verrucomicrobia and contribute to the defense of the ciliate against predators (Petroni et al., 2000); some trypanosomes of the genus *Crithidia* contain intracellular bacteria related to *Bordetella* (β-proteobacteria) with a likely nutritional role in cofactor synthesis (Chang et al., 1975); rhopalodiacean diatoms in nutrient-poor ocean waters are associated with intracellular nitrogen-fixing cyanobacteria closely related to *Cyanothece* spp. (Nakayama et al., 2014); and the intracellular Endomicrobia bacteria in *Trichonympha* and related flagellate protists in termite guts have the capacity to fix nitrogen and synthesize amino acids (Zheng et al., 2015). There is also growing evidence for positive interactions between the unicellular yeast phase of the ascomycete fungus *Candida albicans* and bacteria, especially streptococci, in the oral cavity (Metwalli et al., 2013). (As considered further in section 2.4.2, *Candida albicans* is a dimorphic fungus, i.e., capable of growing as a unicellular yeast or as multicellular filaments.) The streptococci, including species that cause dental caries, provide adhesion sites that promote colonization by the *Candida*, and their fermentative metabolism releases lactate, which is a valuable carbon source for *Candida*, while the consumption of oxygen by the *Candida* reduces oxygen tensions, promoting growth and biofilm formation by the streptococci. These interactions pose important problems for human oral health because the polymicrobial biofilms that includes the streptococci, *Candida*, and other microorganisms increase the resilience of the microbial community to mechanical disruption when we brush our teeth, as well as resistance to antimicrobial agents, including antiseptic mouthwashes.

For other systems, however, the outcome of interactions between protists and bacteria is uncertain or strongly context-dependent. For example, amoebae, e.g., *Acanthamoeba* and *Hartmannella*, are well-documented as reservoirs for important human pathogens, including *Chlamydia*, *Legionella pneumophila*, *Mycobacterium* species, and pathogenic *E. coli* (Molmeret et al., 2005), although whether the amoebae derive benefit or harm from the carriage of these bacteria is unclear. The context-dependence of associations has been studied in detail in an amoeba, *Dictyostelium discoideum*, which is phylogenetically very different from *Acanthamoeba* and *Hartmannella* (figure 2.1C) and has a complex life cycle. *Dictyostelium* alternates between a unicellular stage that feeds on bacteria and a multicellular migratory "slug" which develops into a spore-producing fruiting body. Some *Dictyostelium* strains do not kill all the phagocytosed bacteria, but retain them through the lifecycle, such that amoebae emerging from dispersed spores have a supply of bacterial food. Carriage of the bacteria through the lifecycle is costly for *Dictyostelium*, but this cost is outweighed by the advantage of a food supply under low-nutrient conditions (Brock et al., 2011).

2.4.2. THE FUNGI: HOW MULTICELLULAR RELATIVES OF ANIMALS INTERACT WITH BACTERIA

The evidence for multiple and diverse interactions between protists and bacteria (section 2.4.1) is fully consistent with the proposition that the ancestors of animals interacted extensively with bacteria. However, a more precise test of the proposition is that eukaryotic groups closely related to animals are predisposed to associate with bacteria. The animals belong to a phylogenetically robust supergroup of eukaryotes, known as the Opisthokonta (characterized by the possession of a single posterior flagellum) (figure 2.1C) which also includes a second phylum with a major radiation of multicellular forms, the fungi (Burki, 2014). This provides the opportunity to investigate how bacteria interact with multicellular eukaryotes related to the animals.

Evidence from multiple fungal taxa demonstrate that bacteria can have profound effects on the biology of fungi that form multicellular hyphae and more complex structures, including the basidiomycete toadstools and truffles, and the lichen thallus of fungal symbioses with photosynthetic algae and cyanobacteria. Much of the evidence relates to secondary metabolism. The lichens are renowned for the diversity and complexity of their secondary chemistry, which is generally attributed to the fungal partner. However, there is increasing evidence that the cyanobacterial symbiont *Nostoc* may mediate the synthesis of some lichen substances, including modified amino acids with neurotoxic properties, hepatotoxic microcystin peptides, and microtubule-disrupting polyketides (Kaasalainen et al., 2012). These compounds are believed to provide protection against fungivores.

In some associations between fungi and bacteria, secondary metabolite synthesis is shared between the partners. Research on two different fungal pathogens illustrate this point. One of the fungi is the basidiomycete *Cryptococcus neoformans*, the agent of cryptococcosis in immune-compromised people. The virulence of *C. neoformans* is markedly enhanced by its production of melanin, a black pigment that confers protection against abiotic stressors and host immune attack, but *C. neoformans* in pure culture cannot synthesize melanin. It is dependent on coculture with bacteria, notably *Klebsiella*, which provides the dopamine precursor for melanin synthesis (figure 2.5A). Both the *C. neoformans* and *Klebsiella* obtain protective benefit from their coproduction of melanin (Frases et al., 2006). Dependence on bacteria for secondary metabolite production is also displayed by the zygomycete fungus *Rhizopus microsporus*, the agent of rice seedling blight disease, but, in this system, the fungus interacts with a coevolved, intracellular bacteria of the genus *Burkholderia*, on which it is totally dependent. The bacterial partner has the genetic capacity to synthesize a complex polyketide, rhizoxin, which binds to β-tubulin of the rice plant cells, inhibiting cell division and causing cell death. (The *Rhizopus* β-tubulin lacks the rhizoxin binding site.) Importantly, the rhizoxin is chemically modified by an oxygenase of *Rhizopus*, which inserts an epoxide ring at the C2/C3 position in the molecule (figure 2.5B), resulting in substantially increased toxicity of the rhizoxin to the plant cells (Scherlach et al., 2012).

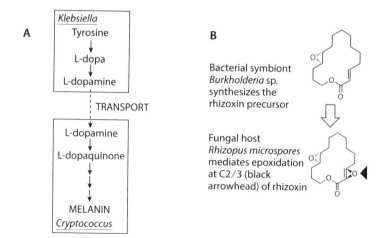

FIG. 2.5. Shared synthesis of secondary metabolites by fungal-bacterial associations. A. Synthesis of melanin by *Klebsiella aerogenes* and *Cryptococcus neoformans*. [Data from Fases et al. (2006).] B. Synthesis of the phytotoxic rhizoxin by the intracellular *Burkholderia* symbiont and its fungal host, *Rhizopus*. (Modified from Fig. 5 of Scherlach et al. [2012].)

Bacterial promotion of secondary metabolite production by fungi is not always mediated by provisioning of metabolic intermediates. A different mechanism, bacterial-induced derepression of fungal gene expression, has been reported in the fungus *Aspergillus nidulans*. *A. nidulans* has the full genetic capacity to synthesize >50 secondary metabolites, but these compounds are not generally synthesized in pure *A. nidulans* cultures. However, in one study *A. nidulans* synthesized various secondary metabolites, including orsellinic acid and lecanoric acid, when it was cocultured with *Streptomyces rapamycinicus*, a common actinobacterium in the natural soil habitat of *A. nidulans*. In these cocultures, the acetylation of histone proteins in *A. nidulans* was markedly increased, including global acetylation of H3K9 and H3K14 (lysine at positions 9 and 14 on histone 3) associated with genes involved in secondary metabolite synthesis (Nutzmann et al., 2011).

There is also evidence that bacteria can have major effects on the developmental biology of fungi. Production of the edible mushroom, the sporophore stage of *Agaricus bisporus*, involves the restructuring of subterranean hyphae to form organized mycelial structure, the primordium, which gives rise to the sporophore (figure 2.6A). This developmental process is not displayed by wild-type *Agaricus bisporus* under standard axenic conditions, but can be induced either by coculturing with bacteria, especially *Pseudomonas putida*, or culturing on axenic activated charcoal (which adsorbs volatile organic compounds, VOCs). There is now good evidence that various VOCs released by the undifferentiated mycelium of *A. bisporus* inhibit differentiation to the primordium and sporophore, and that these compounds are consumed by bacteria (Noble et al., 2009). Developmental decisions by the ascomycete fungus *Candida albicans* are also influenced by bacteria, particularly *Pseudomonas aeruginosa*, which commonly co-occurs with *C. albicans* in the lung of cystic fibrosis patients (figure 2.6B). *C. albicans* is a dimorphic fungus, meaning that it can live as a unicellular yeast or a filamentous form comprising branching hyphae. The filamentous form of *C. albicans* is very susceptible to *P. aeruginosa*, which colonize the hyphae as biofilms that release toxic phenazines and phospholipase C, but the yeast form is resistant. The QS molecules of *P. aeruginosa*, which are acyl homoserine lactones, apparently mimic farnesol, the developmental signal of *C. albicans* that suppresses the signaling pathway mediating the transition from yeast to hyphal growth form, thereby promoting the yeast form that can coexist with *P. aeruginosa* (Hogan et al., 2004).

2.4.3. INTERACTIONS BETWEEN CHOANOFLAGELLATES AND BACTERIA

The choanoflagellates are the closest relatives of the animals. They are continuously in contact with environmental bacteria because they are filter feeders (figure 2.7A), and the interactions between choanoflagellates and

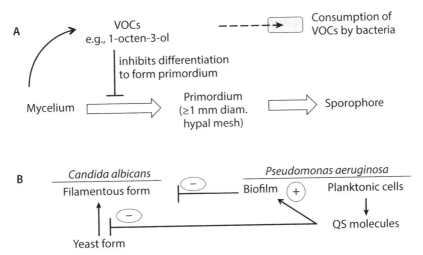

FIG. 2.6. Effect of bacteria of developmental decisions of fungi. A. The developmental transition from undifferentiated mycelium to sporophore primordium in the edible mushroom *Agaricus bisporus* is derepressed by bacteria, e.g., *Pseudomonas putida*, that consume inhibitory VOCs (volatile organic compounds). B. QS molecules of the bacterium *P. aeruginosa* suppress the genes of the fungus *Candida albicans* that mediate the morphological transition from yeast form (which is resistant to *P. aeruginosa*) to the filamentous form, which is particularly susceptible to *P. aeruginosa* biofilms induced by QS system.

environmental bacteria are generally treated as a predator-prey relationship. The extent to which phagocytosed bacteria persist in a viable state within choanoflagellate cells, as commonly occurs for other phagocytic protists (Molmeret et al., 2005) (see section 2.4.1), is unknown.

Recent research on the microbiology of choanoflagellates has focused on interactions between choanoflagellates and food bacteria. Specifically, a subset of food bacteria, including species of the genus *Algoriphagus* (phylum Bacteroidetes), induces the unicellular form of the choanoflagellate *Salpingoeca rosetta* to form distinctive colonies known as rosettes. A rosette comprises a spherical aggregation of one cell and its progeny, linked together by incomplete cytokinesis. This morphogenetic effect is mediated by a specific sulfonolipid, informally known as RIF-1, which is released exclusively from rosette-inducing bacteria (figure 2.7B) and has an extremely low effective concentration of $1-100 \times 10^{-15}$ M (Alegado et al., 2012). Intriguingly, many sulfonolipids and related lipids, including sphingolipids, are bioactive molecules, with multiple instances of sulfonolipid toxins of bacterial origin and sphingolipid signaling molecules that regulate the proliferation, apoptosis, and adhesion of various eukaryotic cells. The functional significance of sulfonolipid-mediated communication between bacteria and choanoflagellate cells is uncertain. Perhaps the rosette morphology modifies water currents to enhance feeding efficiency on certain bacteria, and choanoflagellate cells utilize the sulfonolipids (which could be among-bacterial signaling molecules) as a cue for the presence of these bacteria. Alternatively,

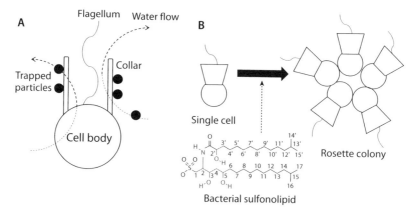

FIG. 2.7. Interactions between choanoflagellates and bacteria. A. The choanoflagellate feeding apparatus comprises a single posterior-directed flagellum, the base of which is surrounded by a collar of interconnected microvilli. The beating flagellum creates an inward-directed current of water, and particles (including planktonic bacterial cells) suspended in the water are trapped on the external surface of the collar and subsequently phagocytosed. B. A sulfonolipid released by bacterial prey induces incomplete cytokinesis of dividing cells of the choanoflagellate *Salpingoeca rosetta*, generating radial arrays of cells known as rosette colonies.

the sulfonolipid may be toxic to the choanoflagellate; and the rosette morphology may be a defensive response that reduces the surface area exposed to the toxin, or a consequence of toxin-induced impairment of the normal cell division process.

Whatever the basis of the interaction between *S. rosetta* and *Algoriphagus* bacteria, the data are fully congruent with information for fungi (section 2.4.2): major morphological transitions in both of these groups of close relatives of animals can be influenced by bacterial products. These findings reinforce the expectation that animals evolved from unicellular eukaryotes that were fully competent in chemical-mediated interactions with bacteria, and raise the possibility that bacterial-derived compounds were, in some way, instrumental in promoting the evolutionary origin of multicellularity, leading to the animals.

2.5. The Animal Condition

2.5.1. THE MICROBIOLOGY OF THE BASAL ANIMALS: THE SPONGES

The basal group of animals is the sponges, phylum Porifera (figure 2.8). The sponge body plan is at the cellular grade of organization, meaning that their cells are not aggregated to form tissues or organs, and it includes <10 cell types. This body plan is fully compatible with hosting a substantial microbial community. A great diversity of microorganisms, including some 30 phyla of bacteria, various Archaea, eukaryotic algae, and other protists, is routinely recovered from sponges (Thomas et al., 2016). In some sponges, the microbiota occupies 1–10% of the total biomass of the association, values similar

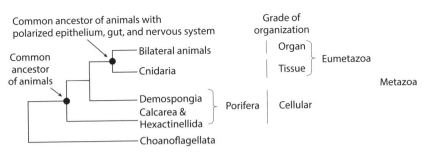

FIG. 2.8. Animal relationships. The animals are the sister group of the phylum Choanoflagellata; the basal animal phylum, the Porifera (sponges) is paraphyletic, with the common ancestor of other animals allied to modern demosponges; and the Cnidaria (jellyfish, corals, etc.) at the tissue grade of organization is sister group to bilateral animals. The phylogenetic relationship between the Cnidaria and other phyla at the tissue grade (Ctenophora [comb jellies] and Placozoa, not shown) is disputed.

to other animals, but the microbiota can account for up to 50% of the biomass in many sponge species (Ribes et al., 2012).

As with choanoflagellates, the sponges come into routine contact with environmental microorganisms because they are filter-feeders. The choanocytes, which are sponge cells morphologically very similar to the choanoflagellates, pump ambient water through the sponge body at very high rates, such that their cells are exposed to a continual stream of water-borne microbial cells. One might expect that the sponge microbiota would be dominated by taxa that are abundant in the surrounding water column and colonize the surface of the sponge cell layers, but this is not the case. The sponge microbiota is taxonomically very different from the microorganisms in the surrounding water and sediment (Thomas et al., 2016). In particular, many of the bacteria identified in sponges (e.g., representatives of the α-proteobacteria, Chloroflexi and Acidobacteria) can be assigned to well-defined clades that are found exclusively or predominantly in sponges. Commonly, the members of a sponge-specific cluster are not structured by sponge phylogeny or location (i.e., the bacteria in sponge species that are closely related or from the same location are not necessarily more similar to each other than those in sponges that are phylogenetically divergent or in different habitats). These results suggest that certain clades of bacteria from various different phyla have broadly based adaptations for the sponge habitat, rather than for specific sponge taxa. In this way, the selectivity of animal-microbiota associations is evident in the simplest of modern animals, consistent with the scenario that the first animals diversified in the context of interactions with resident microorganisms.

What are the traits of the sponge host that dictate which microbial taxa can, and cannot, colonize? In animals generally, the immune system is believed to play a crucial role in controlling the abundance and composition of resident microorganisms, principally by recognizing and eliminating microbial cells that proliferate rapidly or are otherwise deleterious (see chapter 4, section 4.2). So, the first step in exploring how sponges interact with their resident microbiota is to consider whether these basal animals possess an immune system that is recognizably animal-like. Mining the genome sequence of *Amphimedon queenslandica* (*Aq*) (Srivastava et al., 2010) provides valuable information. The elements of the immune system that are of particular interest to understanding how sponges interact with microorganisms are the pattern recognition receptors (PRRs): these are the key microbial recognition proteins of the animal innate immune system (Table 2.1).

TABLE 2.1. The Animal Immune Systems

The innate immune system[1]		The adaptive immune system of vertebrates[4]	
Cellular	Humoral	T cells	B cells
Phagocytosis[2], encapsulation	Recognition of conserved microbial molecular patterns (MAMPs[3]) by animal pattern recognition receptors (PRRs), e.g., Toll-like receptors (TLRs), Nod-like receptors (NLRs), and activation of intracellular signaling cascade, e.g., Toll regulating key transcription factors (e.g., NF-κB) that control the expression of immune-related genes	e.g., cytotoxic, effector, helper, suppressor	Produce soluble antibodies, e.g., IgM, IgG, IgA, IgD, IgE

[1] Defined as lacking memory. However, there is now strong evidence that various invertebrates possess immunological memory, both within the lifespan of an individual animal and transmitted to the next generation, but the underlying mechanisms are not well understood.

[2] In many basal animal groups, phagocytosis is also important for intracellular digestion, and some phagocytic cells can, therefore, function in both nutrient acquisition and defense. In vertebrates, microbial sensors (e.g., mannose-binding lectin, complement factors) initiate complement activation, resulting in opsonization and phagocytosis.

[3] Originally defined as PAMPs (pathogen-associated microbial patterns) (Janeway and Medzhitov, 2002), and subsequently extended to include all microbes (Koropatnick et al., 2004).

[4] Defined as possessing memory. The vertebrate adaptive immunity comprises two types of lymphocytes (T cells and B cells) and is a defensive response linked to a recognition system of exquisite specificity and great diversity, generated by combinatorial rearrangement of gene segments. The genetic changes are strictly somatic, and restricted to the progenitor cells of the two lymphocyte lineages, with the implication that the adaptive immune repertoire is unique to each individual animal, and not inherited via the germline. Lymphocytes can be long-lived and they divide repeatedly when activated, providing the basis for immunological memory. Two different systems have evolved: one based on immunoglobulins (T cell receptors and B cell immunoglobulins) present in all extant jawed vertebrates (shown), and the other based on leucine-rich repeat (LRR) proteins (the variable lymphocyte receptors [VLRs]) in jawless vertebrates. The antigen-binding region of the TCRs and Igs of jawed vertebrates are diversified by random rearrangement of each of variable (V), diversity (D) and/or joining (J) segments of the immunoglobulin gene, and then diversified further by splicing variability and addition of nucleotides during V(D)J assembly. The initial step in the V(D)J rearrangement is a double-stranded DNA break mediated by recombination-activating genes (RAG1/RAG2) acquired horizontally from a bacterium. In the jawless vertebrates (today represented by hagfish and lampreys), the LRRs are assembled in a process akin to gene conversion involving cytosine deaminase genes of intrinsic origin (i.e., not acquired horizontally). The sequence diversity generated by these combinatorial systems far exceeds the number of sequences that can be represented in the lymphocyte populations of an individual animal. The absence of combinatorial immune systems in invertebrates cannot be ascribed to an impoverished genetic capacity because both of the vertebrate adaptive immune systems are based on genes widely distributed in animals. Many invertebrates have immune-related immunoglobulins (e.g., Dscam of insects, FREPs of snails) and LRR-proteins (e.g., Toll receptors), as well as cytidine deaminases; and RAG genes have been identified in the genome of a sea urchin, an invertebrate relative of vertebrates.

In some respects, *A. queenslandica* has a rich PRR repertoire. Its genome includes hundreds of genes with the ancient and conserved scavenger receptor cysteine-rich (SRCR) domain. As in various protists and other animals, many of the *Aq*SRCR genes are predicted to bind to microbial ligands and contribute to immune function; these include sequences that are similar to members of the vertebrate complement system. A second family of PRRs strongly represented in the *A. queenslandica* genome is 135 predicted Nod-like receptor (NLRs) genes (compared to just 22 NLR genes in humans). The NLRs have leucine-rich repeat (LRR) domains predicted to mediate binding to microbial ligands (microbe-assisted molecular patterns, MAMPs) and the NACHT domain for signal transduction (Degnan, 2015) (figure 2.9A). Intriguingly, the sequence of the immunoglobulin (Ig) domains varies widely among the *Aq*NLR genes, and the genome also includes >400 genes with domains implicated in inflammation and cell death (especially CARD and DEATH). This diversity generates the potential for varying the response to different molecular patterns, including discrimination between pathogens and beneficial microbes, although this has not been investigated experimentally. In other respects, however, the PRR repertoire of *A. queenslandica* is impoverished relative to other animals. Most notably, *A. queenslandica* has

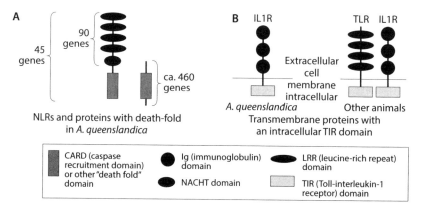

FIG. 2.9. Pattern recognition receptors (PRRs) identified in the genome of the sponge *Amphimedon queenslandica*. A. NLRs (nucleotide oligomerization domain [NOD]-like receptors). Binding of microbial ligands (MAMPs) to the LRR domain induces oligomerization of the NACHT domain, leading to interaction with CARD or other prodeath domains. The resultant signaling cascade culminates in the transfer of a transcription factor of the NF-κB family to the cell nucleus and expression of immune-related genes, especially involved in inflammation and cytokine production. B. Candidate PRRs with an intracellular TIR domain, which mediates signal-transduction to activation of NF-κB transcription factors regulating expression of immune-related genes. It is uncertain whether the extracellular Ig domains of the *A. queenslandica* proteins are ligand binding (Srivastava et al., 2010) or function by interacting with other proteins with extracellular LRR/Ig domains (Hentschel et al., 2012). The *A. queenslandica* genome lacks TLRs with extracellular LRR domains found in other animals.

just two IL1R (interleukin-1 receptor)-like genes, comprising extracellular ligand-binding Ig domains and an intracellular Toll/interleukin-1 receptor (TIR) response domain, which is the defining motif for signaling cascades to activation of NF-κB transcription factors and expression of immune-related genes; and it entirely lacks TLRs (Toll-like receptors), a major family of TIR-containing genes that play very important roles in interactions with microorganisms in other animals, especially insects and vertebrates (figure 2.9B).

To summarize, the data available indicate that the sponges, which comprise the basal animal phylum, have the genetic capacity for sophisticated interactions with microorganisms. Furthermore, the evidence that some gene families contributing to the innate immune system of sponges are conserved, including in mammals and other vertebrates, suggests that the molecular mechanisms of animal-microbial interactions may include common features across the animal kingdom. These conclusions are, however, based on genomic evidence from a single sponge species. As the genomes of other sponge species are sequenced and the experimental study of their immune function expands, we will gain a clearer understanding of how this basal group of animals interacts with its resident microorganisms.

In several important respects, however, interactions between microorganisms and a sponge are different from their interactions with other animals. This is because other animals are more complex: they have more types of differentiated cells than sponges, and their cells are organized into tissues or organs (figure 2.8). We conclude this chapter by considering a key innovation that has had profound implications for interactions with microorganisms: the organization of cells into flat sheets of tissue known as polarized epithelia, including the gut epithelium, found in all animals other than sponges (section 2.5.2).

2.5.2. THE POLARIZED EPITHELIUM AND THE ANIMAL GUT

The polarized epithelium is a sheet of cells that lines the cavities and surfaces of all animals apart from the sponges. As the term suggests, the constituent cells have polarity: the apical surface facing the external surface is functionally distinct from the basal and lateral surface facing internally (figure 2.10A). The cells are bound together laterally by cell junctions, which restricts the movement of large molecules and microorganisms between the cells. The polarized epithelium has two contradictory consequences for resident microorganisms: it creates new habitats for microbial colonization, but it also restricts microbial access to some parts of the animal body.

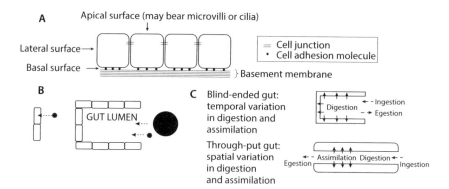

FIG. 2.10. The polarized epithelium and gut of Eumetazoa. A. Organization of the epithelium. The apical surface of many epithelia is a habitat for microorganisms, but the tight adhesion between cells mediated by a belt of cell junctions on the lateral face of the cells forms an effective barrier against inward passage of microbial cells across the epithelium (a simple epithelium [single cell layer] is shown; some epithelia are stratified, i.e., comprise multiple cell layers). B. A gut lumen enables animals to exploit a wide range of food particle size (right), relative to sponges and choanoflagellates which acquire food particles by phagocytosis (left). C. The through-put gut (with mouth and anus) provides spatial variation in function, conferring greater consistency of conditions and resources in each gut region for colonization by microorganisms than in the blind-ended gut, which has very limited spatial variation but considerable temporal variation.

The apical surface of animal epithelia in contact with the external environment is generally colonized by microorganisms. Well-studied examples include the distinctive surface microbiota on *Hydra* (freshwater polyps of the phylum Cnidaria) (Fraune and Bosch, 2007), microorganisms lying between the epithelial cells and external cuticle of most echinoderms, including starfish and sea urchins (Lawrence et al., 2010), and the microbial communities associated with the epithelium of the vagina of female mammals, including humans (Gajer et al., 2012). These communities are taxonomically distinctive, i.e., different from microbial communities in the environment of the animal, and this can be attributed to various extracellular products of the epithelial cells, including polymeric substances (the glycocalyx and, for some cells, also mucus), nutritious metabolites, and antimicrobial factors, that favor certain microorganisms. In this way, the apical surface of animal epithelia creates a novel habitat for microorganisms. But the epithelium also restricts most microorganisms to the apical surface, because it forms an effective barrier against microbial access to internal tissues. The importance of the role of the epithelium in controlling the location of microorganisms is illustrated by the bacterium *Neisseria meningitides*, which is a harmless member of the community colonizing the apical surface of the epithelium of the human upper respiratory tract but can cause life-threatening bacterial meningitis if it breaches the epithelial barrier and infects the meninges (protective membranes bounding the brain and spinal cord) (Rouphael and Stephens, 2012).

In addition to the externally-facing epithelia, many animals have strictly internal epithelia, where the apical surface of the cells face into the body cavity or, as for vertebrate blood vessels, to the blood system. The general consensus is that these epithelial surfaces are microbiologically sterile (or nearly so) in the healthy animal, and this condition can be attributed to the very effective barrier function of surface epithelia, such as the gut, the skin, the respiratory tract and the reproductive tract. There is, however, debate. For example, molecular methods have identified bacterial sequences associated with atherosclerotic plaques adherent to the endothelium (a specialized form of epithelium) lining the arteries in human patients and also the human placenta (which comprises both epithelial and connective layers) at delivery. The argument is whether these results are the product of living cells versus circulating products of dead bacteria or even experimental contamination (Aagaard et al., 2014; Kliman, 2014; Koren et al., 2011), as is considered further in chapter 6 (section 6.4.3).

Of all the various epithelia in animals, the one with the most profound implications for microorganisms is the gut epithelium. The gut is an evolutionary innovation that has enabled animals to exploit food particles larger than the phagocytic limit of cells (ca. 10–30 μm diam.) (figure 2.10B). Because the animal harvests nutritious food items from the environment into its gut, the gut offers nutrient-rich substrates for microbial utilization and, in principle, provides a very favorable habitat for microorganisms. The hazard for the microorganisms is that they may be digested or killed by multiple antimicrobial capabilities, including a very active gut immune system, which protects against microorganisms that may compete for nutrients in the food or attack host cells and tissues. Despite these hazards, many microorganisms can exploit the gut habitat of animals. For example, the lumen of the human gut is estimated to bear ca. 10^{14} microbial cells, weighing approximately 1 kg and accounting for approximately 95% of all microbial cells associated with the human body (see chapter 3, section 3.2).

Animal guts are anatomically and functionally diverse, with implications for interactions with microorganisms. An important distinction is between the blind-ended gut (a cavity) and through-put gut (a tube) (figure 2.10C). In the blind-ended gut, which is ancestral, ingested food is digested by sequential steps, often involving acid, then alkaline, conditions in the gut cavity, prior to absorption of digested nutrients, with minimal spatial variation. The evolution of a through-put gut with spatially ordered differentiation of function, usually from a proximal acidic region, followed by an alkaline region, and distal absorption, was facilitated by the origin of the circulatory system, which mediates the effective translocation of nutrients absorbed from the distal gut throughout the body (figure 2.10C: without a circulatory system, the anterior end of the animal body would be starved of nutrients absorbed from the distal region of a through-put gut). The spatial stability of conditions and resources in each region of the through-put gut is predicted to support microbial communities adapted to these different gut habitats, while the temporal variation in the blind-ended gut creates a highly variable habitat that would be difficult for microorganisms to colonize. If a through-put gut can support stable microbial communities while a blind-ended gut cannot, we would expect the microorganisms in a through-put gut, but not a blind-ended gut, to be taxonomically different from free-living habitat. (We would not expect the blind-ended gut to be microbiologically sterile because microbes are continually being acquired with food.) Consistent with this prediction, a meta-analysis of 16S rRNA gene sequences from different environments discriminated between gut bacteria from animals with a through-put gut (vertebrates, insects, and earthworms) and bacteria from free-living environments, but the bacteria from corals (with a blind-ended gut) were allied to free-living taxa and were possibly dominated by transient forms (Ley et al., 2008). However, this study was not designed to investigate the implications of the evolution of the through-put gut on animal-microbial relations, and further research on the gut microbiota of basal animal groups is much needed. Of particular importance is to obtain gut-specific samples (separate from surface microbiota) and to quantify the abundance and persistence of live microorganisms in blind-ended guts.

2.6. Summary

The ubiquity of interactions between animals and microorganisms is not a trait unique to animals, but an expression of the propensity of all organisms to interact, often to mutual advantage. Even bacteria have a rich social life, mediated principally by the exchange of chemicals (section 2.2). Bacteria share primary metabolites, a process known as syntrophy or cross-feeding, for enhanced access to chemical energy and specific nutrients, and they use chemical compounds as sources of information, enabling them to anticipate or respond to changing conditions.

The predisposition to associate with other organisms is fully integrated into the biology of the eukaryotic cell (section 2.3). All known eukaryotes are derived from a common ancestor bearing *Rickettsia*-like bacteria that evolved into mitochondria. The legacy of the propensity of eukaryotes to interact with bacteria is within every cell of the human body—and of most other eukaryotes. In parallel, unicellular eukaryotes engage in a wide diversity of associations with bacteria, from which they derive nutrients or other benefits, and some protists act as reservoirs for important human pathogens.

The biology of eukaryotic groups most closely related to animals is strongly influenced by bacteria (section 2.4), with evidence, for example, that the sporophore of the edible mushroom cannot develop under microbiologically sterile conditions and that sulfonolipid molecules of bacterial origin determine the morphological organization of choanoflagellates (the sister group of animals).

The antiquity of animal-microbial interactions is further reinforced by genomic research on representatives of basal animal groups, including the sponge *Amphimedon queenslandica*. There had been a general expectation that morphologically-simple animals would have genomes with a small gene content lacking many genes involved in the immune system, which plays a central role in microbial interactions in complex animals. Instead, the genomes of basal animals are gene-rich and include representatives of many gene families involved in immunity (section 2.5.1). These data suggest that the animals have an ancestral molecular capacity to interact with microorganisms. Furthermore, the multicellular condition of the animal creates new and different habitats, including the polarized epithelium and the gut, that microorganisms can colonize (section 2.5.2).

Taken together, these lines of evidence reveal that animals evolved and diversified in the context of preexisting interactions with microorganisms. As the next chapter reveals, these interactions are crucial for the health and fitness of animals, including humans.

3

The Microbiome and Human Health

3.1. Introduction

A major driver of current interest in animal-microbial interactions is the effects of microorganisms on human health. The microbiome has been implicated in virtually every aspect of our lives, from susceptibility to metabolic and atopic disease, to mental health and even the severity of jetlag symptoms. The message is that good health depends on good microbes. This increasing awareness of the microbiology of healthy humans offers great opportunities for improved public health and new therapies, but translating these opportunities into real benefits requires many critical questions to be answered. For example: How can "good microbes" be identified? Are there universal benefits of specific microbial taxa and communities, or do interactions with the microbiome vary with age, sex, and genotype of the human host?

We know more about the microbiome of humans than any other animal species. Detailed catalogs of the microbial taxa and genes associated with humans have been obtained by the endeavors of large consortia. For example, the US Human Microbiome Project (HMP: 2008–2013) quantified the microbiota across 15 body sites and catalogued ca. 1800 bacterial genomes (Human Microbiome Project Consortium, 2012a, b; Ribeiro et al., 2012), and the European Metagenomics of the Human Intestinal Tract (MetaHIT: 2008–2012) determined the microbial taxa and functions in >100 people (Arumugam et al., 2011; Qin et al., 2010). These programs, together with many individual studies, provide a description of the composition of the microbiota, and how it varies with location on the body and among people of different age and health status. From this information, patterns in the composition of the microbiota have been identified, as considered in section 3.2.

A description of the microorganisms associated with humans is necessary, but not sufficient, to understand human-microbial interactions and to apply that understanding for improved human health. We need to understand the processes underlying the microbial effects. Section 3.3 addresses the approaches that are available. Valuable information can be gained from studies on humans, while animal models, especially animals that are microbe-free or have standardized microbial communities, enable experimental analyses that would be unethical or unfeasible for humans.

With the various approaches available, what is known about the effects of the microbiome on human health? Answers to this question are framed by the concept of balance, akin to homeostasis in physiology: processes are regulated to maintain host-microbiome interactions that are healthy for the host, and ill-health is associated with deviation of the host-microbiome interactions from this health state. Most research has focused on the gut microbiota, and section 3.4 addresses the role of perturbation to the gut microbiota in gut disease, cardiovascular disease, and metabolic health.

Finally, there is a perception that this topic has urgency, not only at the scale of the individual patient, but also from the perspective of public health. It is argued that modern lifestyles, especially excessive cleanliness and antibiotic use, are perturbing interactions with microbes, and that these interventions are contributing to the recent increase in various chronic diseases. As considered in section 3.5, these concerns are encapsulated in two complementary hypotheses, the hygiene hypothesis of David Strachan, and the disappearing microbe hypothesis of Martin Blaser.

3.2. The Biogeography of the Human Microbiome

3.2.1. THE DISTRIBUTION AND ABUNDANCE OF MICROORGANISMS

By the best current estimates, >98% of the human microbiota is in the colon (figure 3.1A), which is the distal region of the gut where food material that has not been digested in the stomach and intestine is processed prior to defecation. This means that fecal samples, which are used in most analyses of the human gut microbiota, are representative of the most abundant microbial community in our bodies. Most of the microorganisms in these samples are bacteria, predominantly members of just two phyla, the Firmicutes and Bacteroidetes, together with smaller numbers of other phyla, including Fusobacteria, Actinobacteria, Proteobacteria and Verrucomicrobia (Human Microbiome Project, 2012a). Analyses of the 16S rRNA gene sequences of these bacterial communities consistently reveal high levels of variation within these phyla. A single sample may contain 500–1,000 taxa identified with a 97% sequence identity cutoff (the 97% cutoff is often used as a rough-and-ready index for bacterial species.)

Many of the bacteria in the colon are associated with the protective mucus layer bordering the gut wall. The dominant constituent of animal mucus is mucin glycoproteins (figure 3.2A), and the glycan chains of these molecules provide both adhesion sites and a source of carbohydrate for the colonizing bacteria (Koropatkin et al., 2012). The mucus-associated communities are of particular interest because they have the greatest potential to interact directly with the host, and so influence human health. In the healthy host, the microbial cells in the colon are restricted to the outer part of the mucous layer, and do not penetrate into the inner layer, which is very viscous and adherent to the epithelial cells (figure 3.2B). The importance of the mucus layer is demonstrated by analysis of mice with a null mutation in the gene coding the dominant colon mucin (Muc2$^{-/-}$ mice). In these mice, large numbers of bacteria penetrate to the surface of the colonocytes and even colonize the

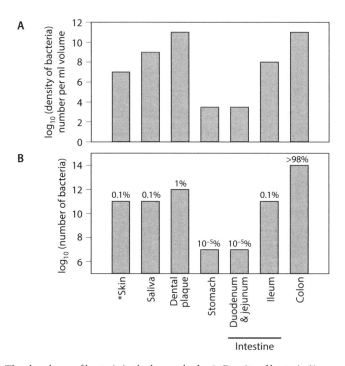

FIG. 3.1. The abundance of bacteria in the human body. A. Density of bacteria (* per cm² skin). B. Total number of bacteria (percent values are percent contribution to total bacterial population in the body). Data are collated from the literature in Table 1 of Sender et al. (2016) and refer to the reference man (25–30 years old, 1.7 m tall, 70 kg weight).

A

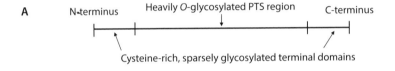

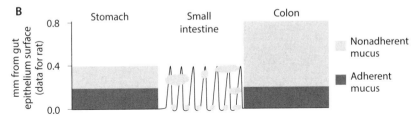

FIG. 3.2. Mucus as a habitat for microorganisms in the gut. A. The dominant component of mucus is mucin glycoproteins, comprising a protein backbone, *O*-glycosylated in the PTS (proline/serine/threonine-rich) region with a great diversity of glycan chains that collectively account for >80% of the mass of the glycoprotein. The mucin monomers are covalently linked by disulfide bridges between cysteine residues at the N and C terminal domains. B. The mucus bordering the stomach and colon comprises an inner adherent layer that is essentially microbe-free and an outer, nonadherent layer that, in the colon, is colonized with many microorganisms. The mucins in the outer layer are partially degraded by microbial glycanases and proteases of both microbial and host origin. The mucus layer in the small intestine is thin and unstratified, and the tips of the microvilli are not always covered. (Redrawn from Fig. 1 of Johansson et al. [2011].)

cells, and the colon of these mice is susceptible to inflammation (Johansson et al., 2011).

Compared to the colon, the stomach and small intestine are hostile environments for microorganisms. The stomach contents are at pH 1.5–3.5 with >0.1 M HCl, and the small intestine, although at a more favorable pH (6.5–7.4 units), contains bile acids and other bacteriocidal secretions. Despite these conditions, the stomach and small intestine contain small populations of taxonomically-distinctive communities of resident microorganisms, as well as microbial cells that have been swallowed from the mouth or transferred by occasional retrograde flow from the colon. The stomach bears yeasts and diverse bacteria, including Firmicutes (especially lactobacilli) and Proteobacteria (Bik et al., 2006); and the microbiota of the small intestine is dominated by Firmicutes bacteria, especially *Streptococcus*, *Veillonella*, *Lactobacillus*, and *Clostridium*, with considerable among-individual variation (Booijink et al., 2010).

The high density of microorganisms in the colon is matched by only one other location in the human body: dental plaque (figure 3.1B). Dental plaque is the biofilm of microorganisms, mostly bacteria, on the surface of teeth and gums. The community is diverse, comprising >700 bacterial species, and the total abundance of microorganisms associated with plaque is greater than the numbers in the stomach and small intestine combined. The microbial communities in the oral cavity are readily accessible to study, revealing remarkable structuring over very small spatial scales (Mark Welch et al., 2016). Following disturbance, for example by tooth-cleaning, the proteinaceous pellicle covering the tooth surface is initially recolonized by *Streptococcus* and *Actinomyces*, which provide a substrate for other bacteria. As the community develops, filamentous corynebacteria play a structurally important role, supporting different bacterial communities at different locations relative to the tooth surface (figure 3.3). Most of the bacteria do not damage the tooth or gum but simple sugars in the food favor acid-producing bacteria, especially *Streptococcus mutans*, resulting in dental caries.

The microbiota of the human skin is taxonomically very diverse, comprising an estimated 1,000 species of bacteria from 19 phyla. A comprehensive metagenomic analysis of multiple skin sites from 15 individuals revealed *Propionibacterium acnes* and *Staphylococcus epidermidis* are very abundant, and comprise multiple strains, with a distribution that varies with position on the body and among individuals (Oh et al., 2014). Fungi account for ≤10% of sequences in most samples, and are dominated by *Malassezia* species, especially *M. globosa* and *M. restricta*. Because these fungi require lipid for growth, they are most abundant in samples taken from areas rich in sebaceous glands, e.g., scalp.

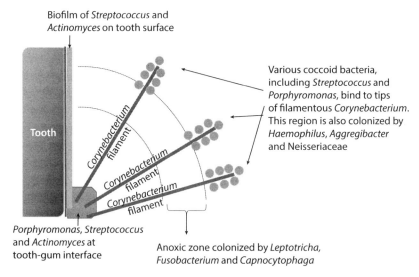

FIG. 3.3. Spatially-structured bacterial communities in dental plaque deposited between the tooth and gum. (Redrawn from Fig. 6 of Mark Welch et al. [2016].)

3.2.2. AMONG-INDIVIDUAL VARIATION AND THE CORE MICROBIOTA

A most surprising result of research on the human microbiome has been that the taxonomic composition of the microbiota varies widely among individuals. This has been demonstrated for all body sites that have been investigated systematically, including the gut, skin, and the vagina (Dethlefsen et al., 2007; Gajer et al., 2012; Oh et al., 2014). Perhaps the greatest of these surprises relate to the microbiota of the vagina. For many years, it was believed that the vagina of healthy women is colonized predominantly by lactobacilli, which maintain a pH<4.5 by their production of lactic acid, and that vaginal communities with low abundance of lactobacilli are predisposed to bacterial vaginosis. This simple perspective has been overturned by detailed longitudinal studies on many healthy women, demonstrating that some women have a stable vaginal microbiota, which may or may not be dominated by lactobacilli, and the microbiota in other women fluctuate between different bacterial communities (Gajer et al., 2012). This variation and its significance are considered from an ecological perspective in chapter 6 (section 6.4.2).

The gut microbiota "can be as unique as a fingerprint" (Dethlefsen et al., 2007). The composition of the microbiota is generally stable in healthy adults over time; in one longitudinal study, 60% of taxa were, on average, retained over a 5-year period (Faith et al., 2013). Unlike fingerprints, however, the microbiota is not absolutely stable, and can change substantially in response to antibiotic treatment (Perez-Cobas et al., 2013), physiological changes such as pregnancy (Koren et al., 2012), and shifts in diet (David, Maurice, et al., 2014). For example, diets with a high meat content tend to favor *Bacteroides* and *Alistipes* (Bacteroidetes), and diets dominated by complex plant carbohydrates favor Firmicutes (e.g., *Roseburia*, Ruminococcus, and *Eubacterium rectale*) and often *Prevotella* (Bacteroidetes) (David, Maurice, et al., 2014; De Filippo et al., 2010).

The demonstration of great among-individual variation in the microbiota has substantial implications for our understanding of the human microbiome. One of the key priorities of large consortium projects on the human microbiome, including HMP and MetaHIT (section 3.1), was to establish whether there is a core microbiota, i.e., a consistent set of microbial partners, and then to investigate the traits of the core taxa and their impact on human health (Hamady and Knight, 2009). It is important to appreciate that the concept of a core microbiota is a way to describe microbial communities and not a testable hypothesis. This is for several reasons. The first is that the core can be defined at any taxonomic level (Shade and Handelsman, 2012). All animals (other than experimentally generated microbe-free individuals) bear microorganisms, and so, logically but not usefully, every animal can be described as having a core microbiota at the highest taxonomic level. In addition, any descriptor of the core microbiota is contingent on the microbiology of all unsampled individuals in the species conforming to the microbiology of sampled individuals, making any claim that a species has a core microbiota provisional. The quality of the microbial inventories also influences the capacity to identify core microorganisms. Some taxa detected in a subset of host individuals with a shallow sampling regime may be discovered to be present in all individuals sampled to a greater depth. For example, in one analysis of human fecal samples, increasing the sequencing depth by 25% yielded a three-fold increase in the number of "core" taxa (Qin et al., 2010). Because the concept of the core microbiota is just a convenient way to describe microbial communities, it is not meaningful, for example, to state that species X has a core microbiota but species Y does not.

When is the concept of the core microbiota useful? As stated in the previous paragraph, it can certainly be valuable to highlight microbial taxa that consistently associate with an animal species. For example, 9 bacterial taxa are reliably recovered from the guts of adult honey bees (Martinson et al., 2011). However, the microbial communities of humans are very complex and variable, making any description of a core microbiota in strictly taxonomic terms unwieldy (Human Microbiome Project Consortium, 2012a). Descriptions of microbial communities associated with humans have tended to shift from consideration of individual species to community types, all members of which may not be detectable in all individuals. Some analyses have partitioned the microbial communities in different people into a limited number of clusters, e.g., six bacterial "community state types" (CSTs) in the vagina of women (Gajer et al., 2012) and three enterotypes of human gut microbial communities (Arumugam et al., 2011). These clusters can be a useful way to reduce the complexity of the system to a few aggregated units, but their validity has been questioned (see chapter 6, section 6.4.2), and their predictive value is uncertain. For example, no association has been identified between a health trait and enterotype in humans.

An alternative route to organize the very large datasets generated by human microbiome studies is to focus on functional traits, and not taxonomic units (Human Microbiome Project Consortium, 2012a). From the perspective of microbial effects on human health, a functional approach is, arguably, more informative than a taxonomic approach because closely related taxa can be functionally distinct, while phylogenetically divergent taxa may have similar functional traits. Metagenomics (i.e., sequencing the total DNA of a microbial community) and metabolomics (quantification of metabolites) are providing valuable information. For example, the microbial genetic capacity for carbohydrate degradation mostly targets simple sugars, such as sucrose and lactose, in the small intestine, but complex glycans in the large intestine (Zoetendal et al., 2012). The predicted consequence is that microbial harvesting of dietary carbohydrate is in competition with the host in the small intestine but augmenting host capabilities in the large intestine.

3.3. How to Study Microbiota Effects on Human Health

3.3.1. EPIDEMIOLOGICAL STUDIES OF THE HUMAN MICROBIOME

Epidemiology is the study of health and disease in populations, including the analysis of environmental factors that may influence health and interventions to improve health. Over the years, epidemiological methods including longitudinal studies (repeated observations of individuals over time) have yielded very important insights, ranging from the association between smoking and lung cancer to the lack of association between childhood vaccines and autism, with important implications for public health policy.

Multiple epidemiological studies are investigating the relationship between the microbiome and human health. One particular focus is the effect of Cesarean delivery (CD) on the health of children, in the light of evidence that the gut of babies born by vaginal delivery (VD) is colonized by microorganisms, especially *Lactobacillus*, from the mother's vagina, while the gut of babies with CD bears bacteria, such as *Staphylococcus* and *Acinetobacter*, derived from skin and the ambient environment (Dominguez-Bello et al., 2010). Does the identity of the first microbial colonists affect human health? So far, there is no clear answer to this question. Various epidemiological data indicate heightened incidence of atopic disease, especially asthma, celiac disease, and possibly type 1 diabetes (T1D) in children born by CD (Neu and Rushing, 2011), but these associations are not found consistently. For example, an analysis of >5,000 children from birth to 8–9 years of age, conducted by the Longitudinal Study of Australian Children (LSAC) program (Robson et al., 2015) did not yield consistent differences between CD and VD children for multiple indices of health, including asthma and social behavior. The children born by CD had a higher body-mass index (BMI) at 8–9 years, but this could be explained by maternal factors, and not the mode of delivery per se. Overweight and obese women are more likely than lean women to have CD, and their children are more likely to have high BMI whether or not the children were born by CD or VD.

The relationship between the gut microbiome and health is also being investigated by an analysis of the incidence of T1D in a cohort of ca. 7,000 children from Finland, Russian Karelia, and Estonia, conducted by the DIABIMMUNE consortium. Although the genetic make-up of the populations in the study area is very similar, the incidence of childhood T1D in Finland is six times greater than in Russian Karelia. As part of the DIABIMMUNE project, a longitudinal study of 33 children at risk of T1D has been conducted (Kostic et al., 2015). The bacterial diversity in fecal samples of children who developed the disease was found to be significantly reduced relative to the children without symptoms, and this reduction was apparent up to a year before disease symptoms were evident. Furthermore, the reduced microbial diversity was associated with indices of metabolic imbalance and gut inflammation (Kostic et al., 2015). The authors of this study raise the possibility that "the T1D-associated microbiota that becomes established prior to disease onset may actively promote a metabolic environment in the gut that is permissive to inflammation and promotes pathogenesis." However, this study cannot demonstrate whether changes in the microbiota affect the onset of T1D—and it was not designed to do so. As is discussed in the following section, different approaches are needed to investigate causation.

3.3.2. CORRELATION, CAUSATION, AND MECHANISM

Most research on the relationship between the microbiome and human health is correlative. This is true for both the large-scale epidemiological studies often involving thousands of people studied over multiple years (section 3.3.1), and also for focused comparisons. For example, fecal samples of patients with irritable bowel syndrome (IBS) tend to have higher levels of *Firmicutes* (especially *Ruminococcus*, *Clostridium*, and *Dorea*) and lower levels of *Bifidobacterium* and *Faecalibacterium* spp. than in healthy volunteers (Rajilic-Stojanovic et al., 2011), and the populations of some *Clostridium* spp., *Akkermansia muciniphila*, *Bacteroides* spp., and *Desulfovibrio* spp. are increased in many patients with type 2 diabetes (Qin et al., 2012). What do these correlations mean? One possible reason for these associations is that certain gut microorganisms contribute to disease symptoms (figure 3.4–1). Alternatively, the difference in microbiota composition between healthy individuals and individuals with a disease could be caused by the disease symptoms (figure 3.4–2), or the two traits (disease symptoms and shift in microbiota composition) could be independent consequences of the physiological dysfunction (figure 3.4–3). A likely example of a correlation without direct causation was described in section 3.3.1: children born by Cesarean section have a greater risk of being overweight than children born by vaginal delivery for the indirect reason that overweight mothers are more likely to have Cesarean sections and the children of overweight mothers are predisposed to be overweight. More complex interactions are also possible. For example, small physiological changes associated with early stages of a disease may trigger changes in the microbiota that reinforce disease symptoms (figure 3.4–4), as suggested by Kostic et al. (2015) for young children with T1D (see section 3.3.1). Just as some microbial communities may exacerbate disease, other microbial communities may reinforce a healthy phenotype.

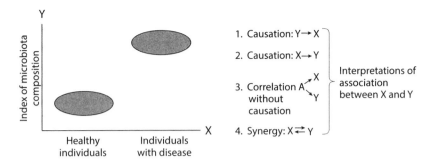

FIG. 3.4. Alternative interpretations of differences in the composition of microbiota (Y-axis) between healthy individuals and individuals with disease (X-axis). These interpretations are discussed in the text at section 3.3.2.

A powerful route to establish causation is to conduct experiments: to perturb one of the correlated traits and quantify the impact on the other trait. Does modification of the microbiota alleviate (or induce) disease symptoms, and does induction (or cure) of the disease alter the microbiota composition? Experiments have been particularly powerful in investigating the relationship between the gut microbiota and obesity. In many (but not all) studies of both humans and mice, the microbiota differs between lean and obese individuals. For example, mice lacking the gene from the 'satiety hormone' leptin (*ob/ob* mice) are obese and differ in the composition of their gut microbiota from wild-type lean mice (+/+). When wild-type mice were colonized with microbiota from the +/+ and *ob/ob* mice, the mice bearing microbes from *ob/ob* mice gained significantly more fat (figure 3.5A), demonstrating that the microbiota in the *ob/ob* mice contributed to the obese phenotype (Turnbaugh et al., 2006).

The experimental demonstration of causation is, however, not sufficient either to understand how the microbiota interacts with human health or to use this information for therapeutic intervention. It is also necessary to establish mechanism, i.e., the sequence of events from composition of microbiota to traits of the animal host or vice versa.

Microbiome research on mammals, including humans, is progressing rapidly from the study of correlation to causation and mechanism. However, researchers studying low diversity microbial communities, especially interactions between the animal and single microbial taxa, have been investigating the problems of causation in animal-microbial interactions for many years, providing useful examples of the relationship between causation and mechanism. In this section, I consider two scenarios with identical patterns of causality, but different mechanisms, using data on insect-microbial interactions.

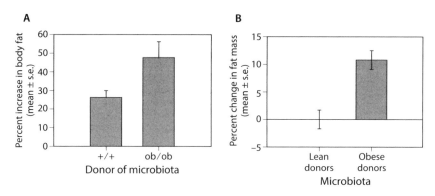

FIG. 3.5. Impact of gut microbiota composition on mouse phenotype. A. The body fat content of mice colonized with gut microbiota from wild-type (+/+) and genetically obese (*ob/ob*) mice. The *ob* gene codes for the "satiety hormone" leptin, a negative regulator of feeding. (Reproduced from Fig. 3C of Turnbaugh et al. [2006].) B. Obesogenic microbiota in obese humans. When mice were colonized with microbiota from twin pairs with discordant BMI, only the microbiota from the obese twin caused increased fat content. (Reproduced from Fig. 1D of Ridaura et al. [2013].)

In the scenario illustrated in figure 3.6A, the microbiota display one trait with downstream consequences on three different host traits. Definitive demonstration of these effects is genuinely difficult in associations involving complex and variable microbial communities, as in the mammalian gut. However, as mentioned above, insights can be gained from relationships involving one or a few microbial partners. Here, I consider a natural association with a single bacterium, *Buchnera aphidicola*, in one group of insects, the aphids. The *Buchnera* bacteria release essential amino acids. Abolition of this single microbial trait limits protein synthesis because the dietary supply of essential amino acids is insufficient, with cascading consequences for multiple physiological traits of the insect. These shifts reduce aphid growth, increase blood osmotic pressure, and drive a shift in the aphid honeydew from watery to viscous consistency. If the underlying mechanisms had not been established, it would be easy to make the erroneous interpretation that the bacteria directly interact with sugar transformations in the gut and host osmoregulatory function. Figure 3.6B shows an alternative route by which microorganisms can influence host traits, where each phenotypic trait of the host is influenced by a different microbial trait. The nutritional relations between *Drosophila* and its gut microbiota, illustrated in figure 3.6B, conform largely to this scenario.

The literature is replete with studies demonstrating correlations between microbiome and host traits without experimental investigation of the direction (if any) of causation, and with untested suggested mechanisms. As the examples in figure 3.6 illustrate, the sequence of events linking correlated traits

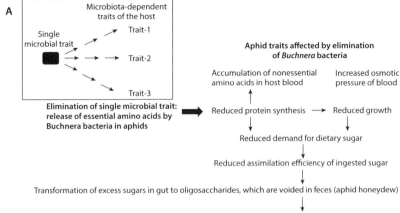

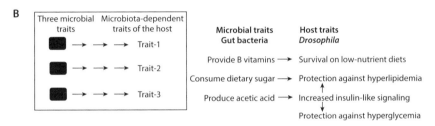

FIG. 3.6. Mechanisms of microbial effects on host phenotypic traits. A. A single microbial trait influences multiple host traits, as illustrated by the effect of essential amino acid release by symbiotic bacteria *Buchnera aphidicola* in aphids (Wilkinson et al., 1997). B. Independent effects of multiple microbial traits on host traits, as illustrated by the impact of gut bacteria on various nutritional traits of *Drosophila* (note that acetic acid production influences two traits, conforming to scenario A) (Shin et al., 2011; Wong et al., 2014; Huang and Douglas, 2015).

can be complex and is not necessarily intuitive. Because different aspects of the biology of animals are interlinked, single or few traits of the microbiota can ripple out to affect many aspects of animal function. Thus, if microbes were found to affect the circadian rhythm of their host, further research is required to investigate whether specific microbial products are interacting directly with the circadian pacemakers of the host (as is considered further in chapter 6, section 6.4.4). Understanding mechanisms enables us to explain causal relationships and to predict the consequences of proposed microbial therapies. Of particular importance to human health, it is only by a thorough understanding of mechanism that one can predict whether a microbial intervention designed to modify one microbe-dependent host trait would also have significant effects on other host traits.

How are we to establish causality and mechanism? Animal models make this possible, as is described in the following section.

3.3.3. ANIMAL MODELS IN HUMAN MICROBIOME RESEARCH

Animal models play a vital role in microbiome research. They are used for experiments to test the principles of microbiome-animal interactions and, specifically in relation to human microbiome studies, to establish causation and mechanism of correlations obtained for humans. The use of animal models is predicated on the assumption of uniformity, or at least similarity, in animal-microbial interactions across different species. This assumption is reasonable because interactions with resident microorganisms are an ancient trait, already well-established in the unicellular ancestors of animals (chapter 2, section 2.4). We should anticipate that, just as for other biological processes, including regulation of the cell cycle, embryo development, and neurobiology, research on microbiota interactions in animal models will shed light on processes in humans.

The principal model animals for human microbiome research are the mouse, zebrafish, and *Drosophila*. These species have the key traits of short generation time, high fecundity, and easy maintenance in the laboratory, and they have attracted a large community of researchers investigating a variety of topics. Many aspects of the biology of these animal models are extremely well-described, sophisticated technologies for genetic transformation and phenotypic screens are routine, and well-maintained databases and large banks of strains and mutants are available as community resources. These many advantages offer a superb foundation for experimental microbiome research. The additional requirement for microbiome studies are that axenic (also known as germ-free) animals can be generated, and colonized with standardized microbial communities, to generate gnotobiotic animals (figure 3.7A). The different models vary in their suitability to answer different experimental questions (figure 3.7B), and, for certain topics, other animal systems can be valuable as an alternative or complement to the three main animal models.

A key advantage of the mouse model is that, as a mammal, it is similar to the human in its overall anatomy and physiology. However, the mouse is a small herbivore, and it has a considerably greater fermentation capacity (represented as a proportionately larger colon and cecum) than the human, and displays coprophagy, which is important for its B vitamin nutrition (Nguyen et al., 2015). Despite these differences between mice and humans, the gut microbiota of the two species have some similarities, with ca. 80 bacterial genera recovered from fecal samples of both the mouse and human, although with different relative abundances (Krych et al., 2013). Germ-free (axenic) mice are technically demanding to produce, requiring Cesarean section under aseptic conditions, and their maintenance involves isolator cages with sterile ventilation and very strict handling procedures. Mitigating

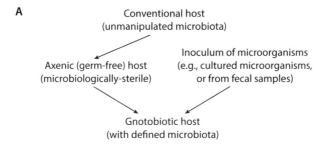

B

Model	Axenic animals		Suitability for large screens & experimental designs	Compatibility with human-derived microbiota
	Ease to generate	Ease to maintain		
Mouse	Technically demanding & costly		Limited by costs	Yes
Zebrafish	Yes	Die at 6–8 days after fertilization	Yes	Aerotolerant taxa only
Drosophila	Yes	Yes	Yes	Aerotolerant taxa only

FIG. 3.7. Animal models for microbiome research. A. Experimental manipulations of the microbiota to investigate microbiome effects on host traits. B. Advantages of different animal models used in microbiome research.

these drawbacks, axenic mice are available from several rodent resource centers and isolators are commercially available. Axenic mice differ from conventional mice in many respects (Smith et al., 2007), some of which are probably downstream consequences, mechanistically far-removed from the microbial-host interaction (see figure 3.6A). Gnotobiotic mice can be generated by gavage (oral administration) of microbial slurries of cultured microorganisms or undefined communities derived from the cecum or fecal samples. Gnotobiotic treatments include the Altered Schaedler Flora (ASF), comprising a defined panel of 8 bacterial taxa that have been isolated from the mouse gut into culture (Dewhirst et al., 1999). The ASF offers total control over the microbiota, enabling standardized experiments over time and across different laboratories (Macpherson and McCoy, 2015), but it does not recapitulate the full complexity of a conventional gut microbiota; for example, ASF-colonization does not recover signature microbial metabolites (e.g., hippurate and chlorogenic acid) in the blood of the conventional mouse (Rohde et al., 2007). Another treatment is humanized gnotobiotic mice, i.e., germ-free mice colonized by microbes from human fecal samples. Great care is needed in interpreting results from these humanized mice because, linked to appreciable differences in microbial composition at the strain-to-species level between mice and humans, many gut microorganisms in humans fail to colonize the mouse (Nguyen et al., 2015; Lagkouvardos et al., 2016). Nevertheless, a causal role for the microbiota can be indicated with some confidence where a human trait (e.g., metabolic or gastrointestinal disease) correlated with a particular microbial community is reproduced in mice colonized with the same microbial community. For example, research on humanized gnotobiotic mice has played an important role in implicating gut microbiota composition in human obesity. In one study of four pairs of human twins discordant for BMI, the microbiota composition differed consistently between the obese and lean individuals, and when the microbial communities were introduced to mice, the mice colonized with the microbiota from the obese humans, but not the lean humans, gained significant amounts of fat (figure 3.5B) (Ridaura et al., 2013).

The zebrafish is emerging as a particularly valuable model for gut microbiota research because the transparency of the larval stage allows rapid microscopical analysis of gut microbiota (especially fluorescently-labeled microorganisms) in real time, and the rapid development rate and small size of the larval fish makes it ideally suited to genetic screens, with classical mutagenesis, RNA interference (RNAi), and reverse genetics technologies well established. Axenic fish are obtained by fertilizing gametes (produced naturally or obtained by expressing from fish) under microbiologically sterile conditions with antibiotics, and the axenic embryos and larvae are reared in isolators or, more conveniently, sterile flasks or dishes. Most experiments involving axenic zebrafish are conducted on unfed larvae up to 6 days post fertilization (dpf) because feeding on sterile food causes lethal epidermal degeneration (the basis of this process has not been determined). Gnotobiotic fish are produced readily by adding microorganisms to the medium. The zebrafish is well-suited to studies of microbial colonization, and the consequences of colonization for gut development, immunity, and nutrition. This can be illustrated by one study on the host immunological response to different members of the gut microbiota. The immune response to colonizing bacteria can be quantified in live fish larvae by the index of numbers of neutrophils (a class of inflammatory immune cells) infiltrating into the gut wall, using a zebrafish strain in which neutrophils are genetically labeled with the green fluorescent protein (GFP). *Vibrio*, which is a dominant member of the gut microbiota in conventional fish, is highly proinflammatory, as indicated by the positive relationship between the numbers of *Vibrio* and neutrophils in the fish guts. By contrast, *Shewanella*, which attains small populations in both conventional fish and in monocolonization, is anti-inflammatory and negatively correlated with neutrophil numbers (Rolig et al., 2015). Intriguingly, fish cocolonized with *Vibrio* and *Shewanella* maintain a low inflammatory status, even though the *Shewanella* account for <10% of the total numbers of bacteria (figure 3.8), demonstrating how low abundance microbial taxa can be functionally important. This large experiment, which used many axenic hosts and exploited the transparency of the fish larvae to score neutrophil numbers, would not have been feasible in the mouse.

The chief advantage of the *Drosophila* model is the ease with which very large numbers of axenic and gnotobiotic individuals can be produced for experiments (figure 3.7B). Surface-contaminating microorganisms are eliminated from hundreds to thousands of eggs at a time by hypochlorite (bleach) treatment, the axenic eggs are then raised on sterile food in vials, and gnotobiotic insects are generated by adding the microorganisms of interest to the food. Combined with the superb genetic and genomic resources for *Drosophila*, these protocols facilitate the genetic dissection of microbiota-dependent traits. For example, studies using multiple *Drosophila* lines with sequenced genomes have revealed considerable host genetic variation in microbiota-dependent effects on host nutrition, including identification of candidate genetic determinants (figure 3.9) (Dobson et al., 2015). Consistent with the prediction made in the opening paragraph of this section that processes mediating host-microbiota interactions are likely conserved across the animals, many of these candidate *Drosophila* genes have homologs

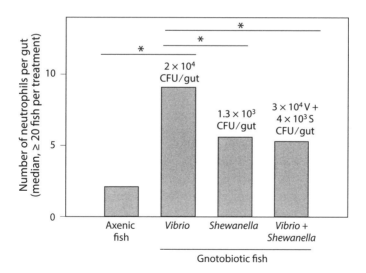

FIG. 3.8. Impact of gut bacteria on neutrophil populations in larval zebrafish. Neutrophil abundance is dictated by the anti-inflammatory *Shewanella* in fish cocolonized with *Vibrio* and *Shewanella* (mean number of CFUs [colony forming units] of bacteria is shown). (Redrawn from Fig. 2E of Rolig et al. [2015].)

across the animal kingdom, including in mammals. As with the zebrafish model, *Drosophila* has very limited value for humanized gnotobiotic experiments (figure 3.7B). The gut microbiota of *Drosophila* is dominated by aerobic and aerotolerant taxa, especially Acetobacteraceae and Lactobacillales, compatible with the oxic environment in the *Drosophila* gut (Wong et al., 2013), and the obligate anaerobic bacteria in the human gut, including many abundant Bacteroidetes and Firmicutes taxa, cannot survive in the *Drosophila* system.

Intriguingly, *Caenorhabditis elegans* has not, to date, been developed fully as a model for microbiome research, despite the successful use of this system to dissect many fundamental aspects of animal biology. *C. elegans* cultures in the laboratory behave as bacterivores, gaining their nutrition from feeding on single bacterial species, usually *E. coli*. However, the apparent dominance of the predator-prey relationship between *C. elegans* and bacteria in the laboratory may not reflect the relationship between *C. elegans* and microorganisms in the natural environment. Wild *C. elegans* bear a diverse gut microbiota (Berg et al., 2016; Felix and Duveau, 2012), raising

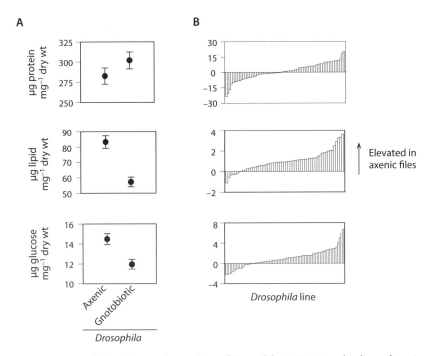

FIG. 3.9. Impact of microbiota on the nutrition of *Drosophila*. A. Nutritional indices of axenic flies and gnotobiotic flies colonized with a standardized microbiota, averaged across 108 fly genotypes. B. Among-genotype variation in response to elimination of the microbiota (Response Index = [difference between mean value of trait in axenic and gnotobiotic flies of each line]/ population mean difference) .(From Fig. 1 of Dobson et al. [2015].)

the possibility that, with appropriate method development, *C. elegans* could become a powerful microbiome model.

Other animal species also have potential as models of the human gut microbiome, even though they lack the extensive resources available for the mouse, zebrafish, and *Drosophila*. From a physiological perspective, the pig is much more similar than rodents to humans, particularly in relation to gastrointestinal disease symptoms, facilitating microbiome interactions with immunological function and enteric pathogens. Protocols for microbiome research, including the development of germ-free piglets, are being developed (Zhang et al., 2013). Among the invertebrates, the cockroach *Shelfordella lateralis* has an omnivorous diet reminiscent of the human diet, can be reared readily in the laboratory, including under axenic conditions, and maintains a complex gut microbiota, including obligate anaerobic bacteria, in its anoxic fermentation chamber. Axenic *S. lateralis* can be colonized successfully with the microbial communities from the mouse cecum, indicating its suitability as a model for mammalian-gut microbiota interactions (Mikaelyan et al., 2016), although humanized gnotobiotic *S. lateralis* have yet to be generated.

3.4. The Microbiota and Human Disease

3.4.1. HOMEOSTASIS AND DYSBIOSIS

The human microbiome is attracting great biomedical interest not because resident microorganisms promote good health, but because they can exacerbate or even cause disease. Current understanding of microbial effects on human health and disease is framed largely by the notion of homeostasis. The physiological systems of animals, including humans, are homeostatic, in that they are stabilized by multiple regulatory processes, especially negative feedback loops, and perturbation of these homeostatic mechanisms results in disease. In an equivalent way, it is argued that interactions between the host and its microbiota stabilize the healthy condition and that perturbation of the interactions results in disease (figure 3.10). The microbial communities associated with disease are often described as unbalanced or dysbiotic.

A substantial body of research on the link between the microbiota and disease has focused on how perturbation of the gut microbiota is associated with disease, especially chronic gut inflammation and cancers, and other physiological systems, including cardiovascular disease and dysfunction of metabolic regulation.

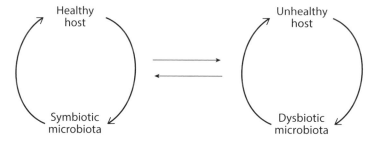

FIG. 3.10. Homeostasis of host-microbiota interactions. Regulatory processes are considered to maintain health-promoting host-microbial interactions or, in a perturbed system, unhealthy interactions with a dysbiotic microbiota.

3.4.2. THE GUT MICROBIOTA AND GUT DISEASE

There is now persuasive evidence that the gut microbiota plays an important causal role in chronic inflammatory diseases of the GI-tract, especially inflammatory bowel disease (IBD, which includes Crohn's disease and ulcerative colitis). This is indicated by the efficacy of antibiotic treatment in ameliorating disease symptoms and by the absence of disease symptoms in rodent models of IBD reared under germ-free conditions. However, IBD differs from the traditional paradigm of diseases caused by microbial pathogens in that there is no persuasive evidence for a single infectious agent. Instead, IBD presents as an immunological disorder, involving heightened activity of proinflammatory processes (e.g., T_H17 cells that release proinflammatory cytokine IL-17) relative to anti-inflammatory processes (e.g., Treg cells which express anti-inflammatory properties, including the production of IL-10), with mutations in immune-related genes (e.g., *NOD2*, *TLRs*) as predisposing factors (Round and Mazmanian, 2009). The apparently contradictory evidence for microbial and immunological determinants of IBD is resolved by the recognition that the microbiota plays an important role in regulating the immunological function in the human gut, as is discussed in chapter 4.3.

A key question to understanding and managing IBD is whether IBD is triggered by specific members of the gut microbiota, especially in individuals with predisposing genotype, or by generalized perturbations to the microbial community (Hold et al., 2014). A role of bacterial pathogens is suggested by the onset or relapse of the disease following acute intestinal infection by bacterial pathogens, including *Campylobacter*, *Klebsiella*, or invasive strains of *E. coli*, while other studies have linked community-level changes, including reduced abundance of Firmicutes, increased Proteobacteria, and decline in the overall diversity of the microbiota. The reduction or loss of specific health-promoting bacteria may underlie the association between reduced microbial diversity and IBD. In particular, *Faecalibacterium prausnitzii* (Firmicutes), which accounts for ca. 5% of the bacterial community in many healthy people, is much reduced in IBD patients and may play a protective role against immunological disorder (Sokol et al., 2008).

Gut microorganisms are also implicated as causative agents of some gut cancers. The role of *Helicobacter pylori* in stomach cancer is well-documented, and there is increasing evidence linking *Fusobacterium nucleatum* with colorectal cancer. *F. nucleatum* is best known as the causative agent of periodontitis (gum infections), and it is generally very rare in human fecal samples. However, it is enriched in some colorectal carcinomas and colonic adenomas (neoplastic lesions that can develop into carcinomas). The causal basis of this association has been demonstrated: *F. nucleatum* promotes colorectal tumorigenesis in the mouse model, where it acts largely by stimulating the recruitment of myeloid immune cells that promote tumor progression and metastasis (Kostic et al., 2013).

3.4.3. THE PERVASIVE EFFECTS OF GUT MICROBIOTA ON HUMAN HEALTH AND DISEASE: CARDIOVASCULAR DISEASE AND METABOLIC SYNDROME

There is now abundant evidence that the effects of the gut microbiota extend beyond the gut to multiple physiological systems and overall health. Some of these effects have been assigned to specific microbe-derived molecules. For example, trimethylamine (TMA), a product of microbial metabolism, has been implicated in the promotion of atherosclerosis (the deposition of fatty material in the inner walls of arteries) and associated cardiovascular disease (Koeth et al., 2013). The gut microbe-derived TMA is converted by flavin monooxygenases in the liver to trimethylamine-*N*-oxide (TMAO), which modifies cholesterol metabolism in multiple ways, all of which promote atherosclerosis (figure 3.11A). These effects are driven by the interaction between gut microbiota and diet, especially L-carnitine, which is a constituent of red meat but absent from vegetables and fruit. An L-carnitine supplement to the diet of laboratory mice caused significant increases in both TMAO levels and the extent of atherosclerotic plaques, but this effect was abolished by antibiotic treatment that suppressed the gut microbiota (figure 3.11B). The relevance of these findings to humans is indicated by elevated TMAO levels in both patients with cardiovascular disease and healthy volunteers administered with oral L-carnitine. Interestingly, the effect of dietary L-carnitine was evident only for omnivorous volunteers, and not long-term vegans/vegetarians (figure 3.11C). The omnivorous volunteers also had higher fasting levels of plasma TMAO (figure 3.11D). These data suggest that the regular ingestion of food containing L-carnitine by omnivores promotes

TMA-producing gut microorganisms. However, this trait has not been associated convincingly with specific members of the gut microbiota, raising the possibility that many taxa may metabolize L-carnitine to TMA.

The diet-dependent effect of gut microbiota on susceptibility to cardiovascular disease is likely one component in a complex web of interactions whereby the gut microbiota influences host physiology. This complexity is revealed by the growing evidence implicating perturbation of the gut microbiota in metabolic syndrome. This is a multifaceted disorder of energy storage and utilization with several linked traits, including obesity, elevated blood pressure, and hyperglycemia, which collectively predispose to type 2 diabetes (T2D), cardiovascular disease, and non-alcoholic fatty liver disease (NAFLD) (figure 3.12A). Although metabolic syndrome is associated most reliably with overnutrition and a sedentary lifestyle, some aspects of the predisposing

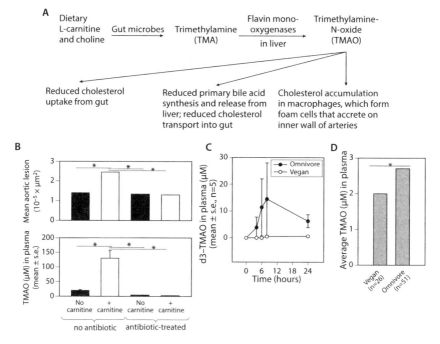

FIG. 3.11. Microbe-mediated effect of dietary L-carnitine on atherosclerosis. A. Production of TMAO and its effect on cholesterol metabolism. Foam cells are fat-laden macrophages, produced when acrophages attack lipid-rich plaques on the artery wall. B. Plasma concentrations of TMAO and area of plaque lesions in the aortic root of *Apoe*$^{-/-}$ mice fed from weaning to 19 weeks on diets with or without L-carnitine. (Redrawn from Fig. 5B and 5C of Koeth et al. [2013].) C. Plasma levels of d3-TMAO in vegan/ vegetarian and omnivore human volunteers over 24 h after oral administration of 250 mg L-carnitine (Redrawn from Fig. 2D of Koeth et al. [2013]). D. Fasting plasma concentrations of TMAO in vegan/vegetarian and omnivore human volunteers. (Redrawn from Fig. 2C of Koeth et al. [2013].)

factors for metabolic disease perturb the microbial communities in the gut, further exacerbating metabolic dysfunction (figure 3.12A). The impact of diet and host metabolic status on the abundance of one bacterial taxon, *Akkermansia muciniphila* (phylum Verrucomicrobia) has attracted particular attention. *A. muciniphila* protects against obesity in rodent models, and its abundance is depressed in both mice fed on high fat diets and in obese humans (Everard et al., 2013).

From a physiological perspective, there are two key traits of the perturbed microbiome in obese rodents and humans: increased extraction of energy from ingested food; and chronic inflammation of organs that regulate metabolism, including the liver, adipose tissue, and muscle. In the following paragraphs, these two traits are considered in turn.

Genetically obese mice and, in some studies, obese people, acquire energy more efficiently from ingested food, as indicated by a reduced energy content of feces. Changes to the microbiota in both the small intestine and colon contribute to this effect (figure 3.12B). The composition of *Lactobacillus* species localized predominantly in the small intestine differ between

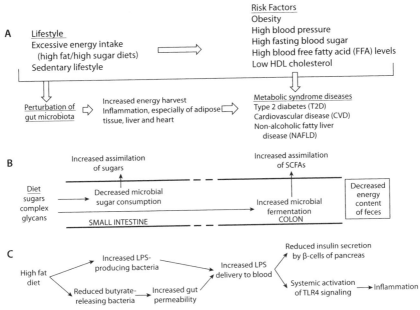

FIG. 3.12. The gut microbiota and metabolic syndrome. A. Perturbation of the gut microbiota by life style and/or risk factors exacerbates metabolic syndrome diseases. B. Changes in carbohydrate metabolism associated with perturbed gut microbiota, promoting energy harvest by the host. C. Metabolic dysfunction associated with increased bacterial LPS (lipopolysaccharides) delivery to the blood and activation of TLR4 (Toll-like receptor 4) signaling causes inflammation and metabolic dysfunction of multiple organs, including the liver, adipose tissue and the insulin-secreting β-cells of the pancreas.

lean and obese people; the abundance of *Lactobacilli* able to utilize simple sugars, especially glucose and fructose, is much lower in samples from obese people than lean people (Drissi et al., 2014), making a higher proportion of ingested sugars available for assimilation by obese people. Compounding this effect, changes in the microbiota of the colon result in increased availability of short chain fatty acids (SCFAs) for assimilation in obese people (Turnbaugh et al., 2009), although conflicting data obtained in different studies prevent definitive identification of the causal taxa. The increased SCFA production also promotes energy harvest by stimulating the secretion of peptide YY, resulting in reduced gut mobility and reduced flux of digesta through the gut, giving more time for digestion and assimilation (Erejuwa et al., 2014).

The chronic, low-level inflammation associated with metabolic syndrome is linked to leakage of LPS (lipopolysaccharide) from gut bacteria into the circulatory system. LPS is a component of the cell wall of Gram-negative bacteria that is highly immunogenic for mammals, including humans. High fat diets promote the abundance of LPS-producing gut bacteria, and these diets also reduce the abundance of bacteria, especially *Eubacterium rectale* and *Clostridium coccoides*, that produce the SCFA butyric acid and promote a strong gut barrier. The combined effect of these changes in microbiota is increased LPS translocation into the circulatory system, and thence to adipocytes (fat cells), liver, and skeletal muscle, where it triggers an inflammatory response (figure 3.12C). In particular, LPS binds to a key receptor in the innate immune system, TLR4 (Toll-like receptor 4), activating a signaling cascade that leads to the release of proinflammatory cytokines, especially IL-1 and IL-6. The TLR4-dependent signaling pathway also suppresses glucose-induced secretion of insulin from β-cells of the pancreas. Mouse studies confirm the causal role of these interactions in metabolic syndrome: mice infused with LPS display insulin resistance and increased adiposity, and these effects were absent from mice either with a null mutation in TLR4 gene or treated with antibiotic to suppress the gut microbiota (Cani et al., 2007; Caricilli et al., 2011).

3.5. The Mass Extinction Event Within?

3.5.1. MICROBIAL COMMUNITIES OF LOW DIVERSITY

As described in section 3.4, various diseases are associated with changes to the composition of microbial communities. Very commonly, these changes involve not only shifts in composition but also reduction in overall diversity. Low microbiota diversity can impair host-microbial interactions in two linked ways. The first relates to functional redundancy (i.e., multiple taxa that perform equivalent functions). Reduced diversity lowers the number of taxa mediating a given function, making sustained function vulnerable to fluctuations in abundance of members of the community. For example, if the number of taxa mediating a function is reduced from ten to two in a community and the different taxa vary in abundance independently of each other, the probability of reduced overall function as a result of stochastic variation in population sizes is considerably greater for the community of two than the community of ten. The second reason for functional impairment in communities of low diversity is that many communities include specific taxa of low abundance but high functional importance. These taxa, which are known as keystone species, are considered further in chapter 6 (section 6.3.2). For now, the important issue is that processes that reduce diversity can result in the chance loss of keystone species, and resultant loss of function.

The loss of microbial diversity can be cumulative over multiple generations. This has been demonstrated experimentally in mice (Sonnenburg et al., 2016). When mice with a humanized microbiota were transferred from a high fiber diet (including corn, soybean, wheat, oats, alfalfa, and beet) to a low fiber diet comprising sucrose and corn meal, the diversity of the gut microbiota declined, and was only partially recovered when the mice were transferred back to the high fiber diet. Over subsequent generations, further reductions in microbial diversity were observed (figure 3.13A), associated with an exacerbated failure to recoup diversity on transfer back to the high fiber diet (figure 3.13B). As the authors comment, "the data demonstrate a diet-induced ratcheting effect" on microbiota diversity.

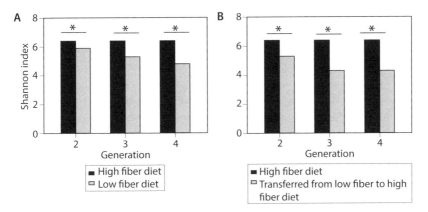

FIG. 3.13. Cross-generation reduction of gut microbiota diversity in mice raised on low fiber diet. A. Mice maintained on high-fiber or low-fiber diet from generation 1 and scored when 5 weeks old in each of generations 2–4. B. Mice maintained on low fiber diet from generation-1 were transferred to high fiber diet when 10 weeks old, and scored when 16 weeks old, with mice maintained on high fiber diet from generation 1 as control. Mean values of Shannon index of bacterial diversity in mouse fecal pellets are shown (5–6 replicate mice per treatment). (Redrawn from Fig. 2B of Sonnenburg et al. [2016].)

This study demonstrates the principle that a low fiber "Western" diet can, over multiple generations, result in an irreversible reduction in microbiota diversity under conditions of extreme cleanliness, as maintained in laboratory mouse facilities. Possible relevance to the human condition comes from the evidence that the diversity of the gut microbiota of humans in industrialized societies is lower than in an agrarian community in Africa (De Filippo et al., 2010) and hunter-gatherer communities in Africa and South America (Clemente et al., 2015; Schnorr et al., 2014). However, these patterns do not provide definitive evidence about the nature of the underlying processes. In other words, we do not have a sufficient basis to conclude that the gut microbiota is undergoing a ratchet-like decline across generations in industrialized human societies.

3.5.2. THE HEALTH CONSEQUENCES OF MICROBIOTA LOSS

An important historical milestone in understanding the beneficial effects of microorganisms for human health came with the Hygiene Hypothesis of David Strachan (Strachan, 1989), who linked the rise in atopic disease (hay fever, asthma, and eczema) in children with small family size and cleanliness. The Hygiene Hypothesis has subsequently been supported by a wealth of epidemiological data, and its scope has been extended from atopic disease to include autoimmune and metabolic diseases (Versini et al., 2015). In parallel, the focus of the discussion has shifted to "the disappearing microbiota" under the argument that the loss of coevolved microorganisms is likely to have more deleterious consequences than reduced exposure to environmental microorganisms (Blaser and Falkow, 2009; Rook and Brunet, 2005).

A strong line of evidence linking the "disappearing microbiota" with poor health outcomes is emerging from epidemiological studies of antibiotic treatments in young children (Cox and Blaser, 2015). For example, one study of children attending a day care center in Finland investigated the association between antibiotic use, microbiota composition of fecal samples, and indices of child health (Korpela et al., 2016). Among the 142 children in this study, the use of macrolide antibiotics was significantly associated with the development of asthma and excess body weight; and the microbial diversity was significantly reduced in the fecal samples of children who had taken antibiotics up to 24 months prior to the analysis, relative to the children who had not taken antibiotics within the previous 24 months.

The epidemiological studies are supported by experimental research, especially on mouse models. Mice administered subtherapeutic doses of penicillin displayed a reduced overall diversity of gut microbiota, including reductions in the abundance of *Lactobacillus*, Rikenellaceae, and segmented filamentous bacteria. Correlated changes in the host included altered expression of genes involved in immune function and increased fat deposition. These effects could be attributed to the antibiotic-mediated changes in the gut microbiota (and not a direct effect of the antibiotic on the mouse) because, in further experiments, the microorganisms in the cecum of the antibiotic-treated mice were introduced to germ-free mice, resulting in a corresponding increase in adiposity of the recipient mice (Cox et al., 2014).

5.4.3. RESTORATION OF THE MICROBIOTA

The accumulating evidence that reduced microbial diversity may be contributing to various chronic immunological and metabolic diseases provides the basis for therapeutic approaches to ameliorate and, in some cases, to cure disease. In principle, the twin approaches of probiotics and prebiotics offer great promise to restore the microbiota, with consequent health benefits. Probiotics refers to the "live microorganisms which when administered in adequate amounts confer a health benefit on the host" (Hill et al., 2014); and prebiotics comprise dietary supplements of complex glycans that can be utilized by specific beneficial microbes in the gut, so supporting and often amplifying their populations.

Probiotics and prebiotics are widely available in fortified foods (e.g., yoghurts) and as food supplements. The diversity of probiotic bacteria and prebiotic glycans used in these preparations is generally small, dominated by *Lactobacillus* and *Bifidobacterium*, and by inulin (fructose polysaccharide), galactooligosaccharides, acacia gum, and psyllium, respectively; and they cannot replicate the tremendous diversity of microorganisms in the microbiota of a healthy person or diversity of plant glycans in a balanced diet. Foods and food supplements with probiotic or prebiotic properties are being consumed very extensively by the general population, despite a dearth of clinical trials on their health value. Although there is a widespread view that probiotic and prebiotic supplements can be beneficial for individuals with metabolic disease, chronic inflammatory diseases (e.g., IBD, atopy) or chronic infections, e.g., vaginosis, (MacPhee et al., 2010; Mekkes et al., 2014; Whelan and Quigley, 2013) and are, at worst, harmless for healthy people, there are also counter-indications, with some suggestions that these interventions may have pro-obesity effects (Angelakis et al., 2013). The variability in reported outcomes should not be a surprise because the composition of the resident microbiota and the genotype and physiological condition of the host are all predicted to influence the effect of newly introduced taxa or novel dietary constituents.

Microbial therapies are also being applied to resolve pathogenic infections. Remarkable success has been achieved for fecal microbiota transplantation (FMT, i.e., the delivery of fecal material from a donor to the patient) to resolve *Clostridium difficile* infections (Debast et al., 2014). Because FMT involves the administration of an undefined cocktail of microorganisms, its efficacy can be variable, creating a strong incentive to develop microbiologically standardized protocols. The demonstration that clostridia including *Clostridium scindens* confers resistance to the intestinal pathogen *Clostridium difficile* in both human patients and mouse models, offers a way forward toward standardization. The beneficial effect of *C. scindens* has been attributed to its unusual pattern of metabolism of bile acids, with products (deoxycholate and lithocholate) that are toxic to *C. difficile* (Buffie et al., 2015).

3.6 Summary

Biomedical science and the prospect of microbial therapies for a range of chronic diseases are a major driver of microbiome science. We have more information on the microbiome of humans than any other animal, including extensive catalogs of taxonomically important sequences, especially bacterial 16S, metagenome data and genome sequences of individual microbial taxa (section 3.2). This information is the basis for comprehensive characterization of the human microbiome and its variation. Most of the microorganisms in the human body are in the large intestine (colon), and their computed weight is roughly equivalent to the weight of the human brain. The composition of the microbial communities associated with humans varies widely with location in the body and, within each location, among individual people. Some of the variation can be related to age, lifestyle, health status, and ethnicity, but much of the variation appears to be idiosyncratic.

Epidemiological studies on the human microbiome are revealing correlations between microbiome traits and health, especially in relation to metabolic and immune-related disease, and these approaches are being complemented by experimental studies on model organisms to understand the causal basis of these correlations (section 3.3). The laboratory mouse is a particularly valuable tool for research on the gut microbiota. Germ-free mice can be "humanized", i.e., colonized by microorganisms from human fecal samples, to test for causality and investigate mechanism. Other useful model systems are the larval zebrafish, where experimental studies are facilitated by the transparency of its body and gut, and *Drosophila* fruit flies, which are particularly amenable to large-scale experimental designs and mechanistic studies.

Combined studies on humans and model systems have shown how perturbations of the microbiome (a condition known as dysbiosis) can cause or exacerbate disease (section 3.4). For example, obesity is commonly linked with reduced populations of *Akkermansia muciniphila*, inflammatory bowel disease is correlated with low abundance or loss of *Faecalibacterium prausnitzii*, and *Fusobacterium nucleatum* is positively associated with some colorectal cancers.

The gut microbiome can also have health effects at a distance, including predisposition for atherosclerosis of the arteries and metabolic syndrome. Western lifestyles, including diet, antibiotic use, and excessive cleanliness, are widely argued to reduce the diversity of the human microbiome, and the resultant dysbiosis may be associated with various chronic diseases (section 3.5). Consistent with this scenario, high-fat/low-fiber diet administered over several generations results in a net reduction of diversity in the gut microbiota of mice. Probiotics and fecal microbiota transplants are being applied for microbiome restoration and, although their use is exceeding demonstrations of efficacy, some positive indications of their value are emerging.

4

Defining the Rules of Engagement

THE MICROBIOME AND THE ANIMAL IMMUNE SYSTEM

4.1. Introduction

Traditional understanding of the immune system is couched strictly in antagonistic terms: the immune system recognizes molecular patterns indicative of nonself, including those associated with microorganisms, and mounts a defense response that eliminates the foreign entity. The effectiveness of the defenses mounted by living animals is vividly demonstrated by the speed with which a cadaver is colonized and consumed by microorganisms, with the early stages of dissolution mediated partly by microorganisms that had coexisted in apparent harmony with the living animal (Metcalf et al., 2016). The evidence that the immune system is central to the defensive capability of animals is overwhelming. Animal models (mouse, *Drosophila*, *C. elegans*) with specific genetic lesions in immune function suffer septicemia and early death, and for humans heightened susceptibility to opportunistic infections is a major cause of morbidity and mortality in immunocompromised patients.

The interpretation of the immune system as a protector of "self against all comers" is enormously powerful. It has underpinned dramatic medical advances, ranging from vaccination to treatments for allergies and the promise of cancer immunotherapies. Nevertheless, this concept of animal immunity has limitations. Most notably, it explains the colonization of animals by microorganisms exclusively in terms of microbial suppression or evasion of host defenses. Although these processes undoubtedly occur, they are unlikely to provide a complete explanation for the microbiology of animals. Alternative pardigms have been developed. For example, the immune system can be envisaged to detect and respond to molecular indices of danger (Matzinger, 2002), and the definition of immune function can be expanded to include immunological tolerance, i.e., the processes that limit the negative consequences of colonization by microorganisms (Ayres and Schneider, 2012). By these perspectives, the persistence of certain resident microorganisms can be attributed to immunological tolerance of microorganisms that do not generate danger signals. The animal immune system would be analogous to a fighting unit with clearly defined rules of engagement to avoid attacks on the civilian population of benign or beneficial microorganisms. Although superficially appealing, this interpretation that the microbiome is essentially isolated from the immune system appears to be wrong. The resident microorganisms are not bystanders in the animal conflict with pathogens and parasites, but play an important role in immune function of the healthy animal.

The extensive and multifaceted engagement between the immune system and microbiota is fully consistent with the recognition that the animal immune system has evolved and diversified in the context of long-standing and pervasive interactions with resident microorganisms—and that the evolving immune system has provided a further molecular and cellular arena with which the microbiota has engaged (chapter 2: section 2.5 and table 2.1). The interactions between the immune system and the microbiota is a young discipline, and the rules of engagement between the immune system and the microbiota are far from understood. Even so, the interactions can conveniently be considered as the reciprocal processes of how the immune system regulates the microbiota and, in reverse, how the microbiota regulates and complements the immune function of the animal. Section 4.2 describes the evidence that the abundance, distribution, and composition of the microbiota is controlled through tightly regulated spatiotemporal variation in both antimicrobial immune molecules and immune cells, such that the animal habitat is a patchwork of permitted and forbidden territories for different microorganisms. Section 4.3 addresses how members of the microbiota affect the production and activity of many immune effectors, modulating host immunological determinants of their abundance and distribution. There is now accumulating evidence that the immunological health of the animal is dependent on these microbial activities. In addition to these modulatory effects of the microbiome on the animal immune system, certain microorganisms in many animals provide defensive functions, augmenting the animal's defensive capabilities against pathogens and other natural enemies (section 4.4).

4.2. Immune Effectors and the Regulation of the Microbiota

4.2.1. IMMUNOLOGICAL CONTROLS OVER THE ABUNDANCE AND DISTRIBUTION OF MICROORGANISMS

As the poet Robert Frost famously wrote: "good fences make good neighbors." The fences that assure coexistence between an animal and its microbiota are predominantly immunological. (The organization of the animal immune system, including the distinction between adaptive and innate immunity, is described in chapter 2, Table 2.1). The central role of the immune system in regulating the microbiota is illustrated most vividly by the relationship between the mammalian gut and its resident microorganisms. Virtually none of the 10^{14} microbial cells in the gut lumen of a healthy human gain access to the internal organs. The restriction of microorganisms to the gut lumen can be attributed largely to immune effectors acting in concert with key cellular processes, including close adherence between the gut epithelial cells and continual sloughing of mucus and cells from the gut wall surface.

In mammals, immunoglobulin A (IgA) plays a central role in restricting the abundance and distribution of gut microorganisms. IgA is the dominant class of antibodies produced in the gut mucosa, and many IgA molecules have high affinity for epitopes displayed by gut microorganisms. IgA is secreted onto the apical surface of epithelial cells and into the mucus, where its coats microbial cells, preventing the microorganisms from making contact with the gut epithelium or penetrating to the internal tissues of the animal. Overall, the molecular diversity of the antigen-binding sites of the IgAs is very high and dynamic, varying with location in the gut and over time in any one gut region.

The mechanism by which microbiota-specific IgAs are produced is understood in outline (figure 4.1) (Macpherson and Uhr, 2004). Microbial cells and soluble antigens that make contact with the epithelial surface are taken up by specialized immune cells, including dendritic cells, associated with the epithelial surface and presented to lymphocytes in gut-associated lymphoid tissues, resulting in T cell-dependent induction of IgA-producing B cells. These activated B cells are then distributed widely in the gut mucosa, and the IgA that they secrete is translocated across the gut epithelium to the luminal surface. The IgA-secreting B cells do not, however, spread away from the gut to other organs, and so a systemic immune response to antigens of the gut microbiota is not generated. Mice that lack the capacity to produce IgA have an altered gut microbiota, including increased abundance of segmented filamentous bacteria and other anaerobic Firmicutes, together with heightened susceptibility to infection and activation of the systemic immune system arising from escape of bacteria into the body tissues (Suzuki et al., 2004).

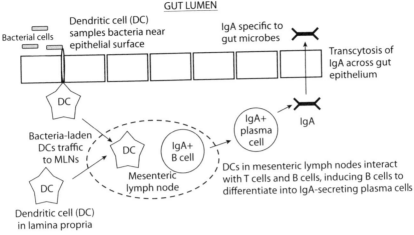

FIG. 4.1. Production of immunoglobulin A (IgA) molecules specific to gut microorganisms. IgA contributes to the restricted access of gut microorganisms to the gut epithelium. MLN: mesenteric lymph node.

The function of the adaptive immune system in limiting the number and distribution of the gut microorganisms is complemented by multiple innate immune effectors, including lectins, antimicrobial peptides (AMPs), and various enzymes that produce reactive oxygen species (ROS, e.g., peroxide, hydroxyl radicals, etc.) and reactive nitrogen species (RNS, e.g., nitric oxide). Innate immune effectors are generally less specific than IgAs because they are active against conserved traits displayed by many microorganisms, such as the peptidoglycan of the bacterial cell wall and the β-1,3-glucans of fungal cell walls. These conserved molecular traits were originally described as pathogen-associated molecular patterns (PAMPs) but, with the recognition of the ubiquity of nonpathogenic microorganisms, the "pathogen" of the PAMP is increasingly generalized as "microbe" (MAMPs) (see Table 2.1). For example, bacterial peptidoglycan is a MAMP that is recognized by a C-type lectin, RegIIIγ, which is secreted from epithelial cells of the small intestine in the mouse and accumulates in the mucus adjacent to the epithelium (Vaishnava et al., 2011). RegIIIγ binds strongly to bacterial peptidoglycan, resulting in cell wall disruption and death of the bacterium, thus limiting bacteria to the outer (luminal) portion of the mucus. Mice that lack RegIIIγ bear dense populations of bacteria that penetrate to the intestinal epithelial surface (figure 4.2A) and activate the adaptive immune system, as indicated by an increase in IgA-producing immune cells and elevated levels of IgA in the feces (figure 4.2B).

Further insight into the role of the immune system in regulation of the microbiota comes from research on insects, especially of *Drosophila* and mosquitoes. In these insects, a critical role is played by the dual oxidase (DUOX) enzyme expressed in the midgut. DUOX mediates the production of multiple

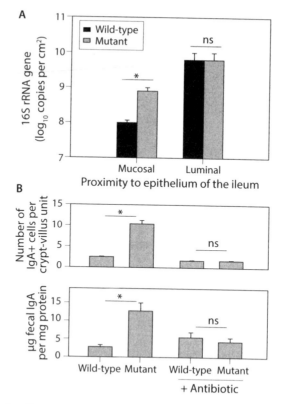

FIG. 4.2. Effect of a null mutation in the gene *RegIIIγ* on the gut microbiota in the mouse ileum. A. Abundance of bacteria quantified as 16S rRNA gene copies in the ileum. (Redrawn from Fig. 3D of Vaishnava et al. [2011].) B. Abundance of IgA-producing cells in the ileum and IgA in the feces of the mice. (Redrawn from Fig. 4B&C of Vaishnava et al. [2011].)

ROS, including superoxide, peroxide, and hypochlorite, via two enzymatic activities, NADPH oxidase (NOX) and peroxidase (figure 4.3A). The midgut DUOX of *Drosophila* suppresses the populations of gut microorganisms partly by direct toxic effects on the microbial cells, but also by increasing the turnover of gut epithelial cells, which makes the gut environment less stable for the microorganisms (Kim and Lee 2014). In young adult mosquitoes, the midgut bacterial population is very small, but increases dramatically when the adult female ingests a bloodmeal (Oliveira et al., 2011), and this effect is associated with a dramatic reduction in ROS in the midgut (figure 4.3B). Evidence that this negative relationship between ROS and bacterial abundance is mediated by the activity of the midgut DUOX comes from RNAi experiments, showing that reduced expression of the DUOX gene causes an increase in bacterial populations in sugar-fed insects (figure 4.3C).

The immune system also plays an important role in regulating the abundance and location of certain intracellular microorganisms. For example, the weevil

Sitophilus bears a γ-proteobacterium *Sodalis pierantonius* within specialized cells called bacteriocytes; and the bacteriocytes display enriched expression of just one known antimicrobial peptide, coleoptericin-A (ColA). When *E. coli* or other Gram-negative bacteria are treated with ColA, they adopt the striking phenotype of polyploidy and exceptionally large cell size (up to 20 μm long), which is also displayed by the *S. pierantonius* cells in the bacteriocytes. When expression of the weevil *colA* gene is reduced by RNAi, the *S. pierantonius* populations become dominated by small cells with low genome copy number, some of which are found in tissues beyond the bacteriocyte, indicating that ColA functions to suppress proliferation of the bacterial symbiont and prevent its spread beyond the bacteriocyte (Login et al., 2011).

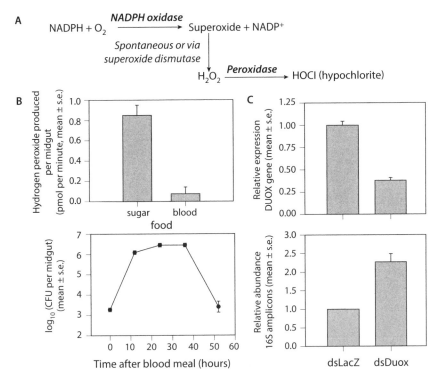

FIG. 4.3. Regulation of midgut bacteria by dual oxidase (DUOX) in *Aedes aegypti* mosquitoes. A. The two enzyme functions of DUOX (NADPH oxidase and peroxidase, in bold). B. Blood-feeding by adult female mosquitoes for 24 h suppresses the capacity of the midgut to generate hydrogen peroxide and increases the bacterial population in the midgut over 36 h. (Redrawn from Oliveira et al. [2011] Fig. 1H and Fig. 5A.). C. RNAi against *DUOX* gene (dsDUOX) reduces expression of the *DUOX* gene in the mosquito midgut and increases the midgut bacterial population, as quantified by 16S rRNA gene amplicon abundance. In this experiment, 2-day-old female mosquitoes were injected with dsDUOX or the negative control dsLacZ, and assayed 48 h later. (Redrawn from Oliveira et al. [2011] Fig. 7C and 7D.)

4.2.2. IMMUNOLOGICAL REGULATION OF THE COMPOSITION
OF THE MICROBIOTA

So far, we have considered how the animal immune system limits the abundance and distribution of microorganisms. However, microbial taxa vary in their susceptibility to the immune responses mounted by the animal

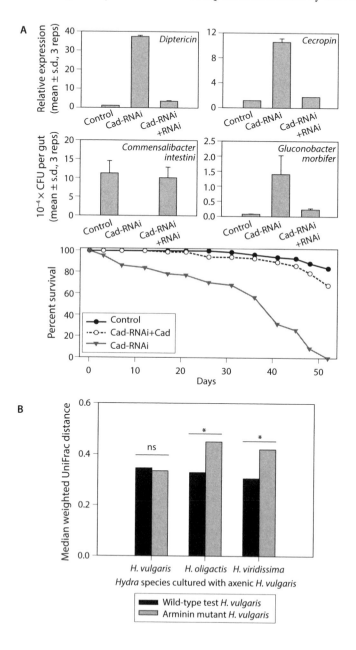

host, with the consequence that the host immune system can additionally influence the composition of the microbiota. Three invertebrate systems illustrate this principle.

The first example is the relationship between the bobtail squid *Euprymna scolopes* and bioluminescent bacterium *Vibrio fischeri*. The juvenile squid acquires an inoculum of bacteria from the water column by trapping bacterial cells in mucus strings that are then transported via cilia into pores which lead, via a duct, to the light organ. The mucus and duct have high levels of nitric oxide and other RNS, as well as ROS and possibly other humoral immune effectors that collectively kill most microbes; but the native symbiont, *Vibrio fischeri*, is remarkably resistant to this immunological attack, resulting in unibacterial (and often clonal) colonization of the light organ (Davidson et al., 2004). The balance of evidence indicates that the ROS/RNS production is a generalized antimicrobial response, and that the remarkable stress-resistance and detoxification capacity of the native symbiont is required to gain access to the light organ. ROS/RNS-tolerant bacterial pathogens could exploit this entry route to the squid, and other host mechanisms are predicted to select specifically for the native symbiont.

Beneficial microbial symbionts are not, however, always the most resistant to host antimicrobial functions. This is illustrated by research on the gut bacterial communities in one laboratory strain of *Drosophila*. The microbiota in unmanipulated flies includes a dominant beneficial bacterium, *Commensalibacter intestini*, and low numbers of a related bacterium, *Gluconobacter morbifer*. When the expression of AMPs in the *Drosophila* gut was stimulated by genetic dysregulation of gut transcriptional factors, the populations of *C. intestini* declined, and, in the absence of these protective bacteria, *G. morbifer* increased dramatically, causing high fly mortality (figure 4.4A) (Ryu et al., 2008). In this system, dampening elements of

FIG. 4.4. (Opposite page) Structuring of the microbial community by antimicrobial peptides (AMPs). A. In *Drosophila*, RNAi-mediated suppression of expression of the gene caudal (*Cad*, coding a transcription factor) promotes expression of the AMPs *diptericin* and *cecropin* (top), with consequent changes to the abundance of gut bacteria *Commensalibacter intestini* and *Gluconobacter morbifer* (middle) and survival of *Drosophila* (bottom). The *Cad*-RNAi line is compared to the parental line (control) and genetic reintroduction of *Cad* (*Cad*-RNAi+*Cad*). (Redrawn from Fig. 5B, C and E of Ryu et al. [2008].) B. Arminin, an AMP of *Hydra vulgaris*, promotes bacterial communities that resemble the native community. Test *Hydra vulgaris* (comprising axenic animals that were either wild-type or null mutants for the AMP arminin) were co-incubated with conventional *Hydra* of the same or different species for 5 weeks, when the bacterial communities associated with the test *Hydra* were quantified. The weighted UniFrac distance is a measure of the difference between the composition of the bacterial communities associated with the test *Hydra* and the native community on *Hydra vulgaris*. (Redrawn from Fig. 5E of Franzenburg, Walter, et al. [2013].)

the host immune system is essential for maintenance of the beneficial microbiota. Furthermore, the deleterious consequences of overexpression of antimicrobial peptides in the *Drosophila* gut relate to dysfunction of the gut microbiota and not the more widely recognized trade-offs of heightened immunological activity with self-damage or resource depletion (Sadd and Schmid-Hempel, 2009).

Immune effectors also play a critical role in determining the composition of the microbiota in *Hydra*, a morphologically simple animal comprising just two epithelial layers. The composition of bacterial communities associated with *Hydra* differs among *Hydra* species and is very stable within one species. For example, the same species-specific signatures of the communities in *H. oligactis* and *H. vulgaris* are found in wild *Hydra* populations collected from fresh water lakes and in cultures of *Hydra* maintained in the laboratory for at least 13 years (Fraune and Bosch, 2007). When germ-free *Hydra vulgaris* were incubated with unmanipulated individuals of other *Hydra* species, they were colonized by bacteria, and the composition of the bacterial communities was more similar to *H. vulgaris* than the donor *Hydra* species (figure 4.4B). However, this capacity of germ-free *H. vulgaris* to select its species-specific microbiota was significantly impaired in *H. vulgaris* mutants that were unable to produce a specific family of antimicrobial peptides with bacteriocidal function, known as arminins (Franzenburg, Walter, et al., 2013). In this way, the arminin profile dictates the composition, as well as the abundance and location, of the bacteria associated with the *Hydra* epithelial cell surface.

The required role of arminins to generate the species-specific microbiota composition in *Hydra* raises an important question of evolutionary mechanism. One scenario raised by Franzenburg, Walter, et al. (2013) is that the arminin profile of *H. vulgaris* evolved to promote the establishment of the species-specific microbiota. This interpretation expands the definition of immune effectors beyond a defensive role, i.e., the immune system functions to reduce microbial abundance, tissue distribution, etc. One possible scenario for the evolution of immune effectors as positive regulators is shown in figure 4.5. A microbial partner may use a host immune effector as a reliable cue that microbial competitors are depleted or cleared from the habitat and that the habitat is suitable for colonization; and an immune effector may, through host-microbial coevolution, become a trans-kingdom signal that promotes microbial colonization and persistence. However, the data of Franzenburg, Walter, et al. (2013) are fully compatible with the traditional interpretation of immune effectors with a strictly defensive function: the specificity of the arminin has been selected by antagonistic interactions with unknown pathogens yielding a species-specific microbiota that happens to be resistant or tolerant. Further

1   Pathogen-induced immune response
    is not deleterious to mutualist

⇩

2   Pathogen-induced immune factor is
    information molecule for mutualist,
    e.g., altered proliferation, tissue tropism,
    functional traits

⇩

3   Immune factor is produced constitutively
    and functions as cue or signal that
    regulates mutualist function

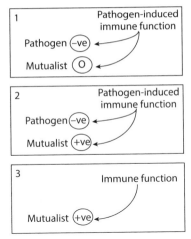

FIG. 4.5. Scenario for the evolution of immune factors as positive regulators of beneficial microorganisms (mutualists). Definitive evidence for this scenario is lacking (see text for details).

research is required to assess the incidence and significance of the evolutionary transition of immune effectors as negative and positive regulators of the microbiota, especially bearing in mind the growing evidence that antimicrobial peptides can serve multiple functions beyond killing microbial cells, including immune modulation and wound healing (Rolff and Schmid-Hempel, 2016).

4.2.3. THE INTERSECTION BETWEEN HOST IMMUNOLOGICAL FACTORS AND NUTRIENT SUPPLY TO THE MICROBIOTA

The microbiota of an animal is not shaped by immune effectors alone. Multiple other factors have been implicated, including diet and hormonal titers, and the genotype, sex, and age of the host. Although these many determinants of the microbiota are not independent, their relative importance and how they interact are poorly known.

An intriguing demonstration of how the immune system can dictate nutrient availability for resident microorganisms, and consequently influence the abundance and composition of the microbiota, comes from research on the carbohydrate composition of mucins and other glycoproteins released from mammalian gut epithelial cells into the gut lumen. The structure of mucins is described in chapter 3 (see figure 3.2A). Early analyses revealed a remarkable difference in the chemistry of the mucins between conventional mice and germ-free mice: in the conventional mice, the terminal sugar unit of the glycans decorating the mucins is predominantly fucose, but in germ-free mice it is mostly sialic acid. The reason for this difference is that gut bacteria, including the common species *Bacteroides thetaiotaomicron* (abbreviated to *B. theta*), stimulate the expression of a host gene *FUT2* coding a fucosyltransferase in the gut epithelial cells (Xu et al., 2003). The fucose-dominated glycoproteins are advantageous to *B. theta*, which possesses a fucosidase that cleaves the terminal fucose residues, together with the transporters and metabolic enzymes required to internalize the fucose and utilize it as a carbon source (figure 4.6). In this way, *B. theta* gains access to a carbon source, enabling it to outcompete other bacteria that cannot utilize fucose.

The most parsimonious route mediating the production of fucose-dominated glycoproteins in mice bearing gut microbes would be induction of the host FUT2 by a bacterial product, but this does not occur. Instead, the *FUT2* gene in the gut epithelial cells is regulated by the immune system, specifically the cytokine IL-22, released by a subgroup of innate lymphoid cells known as ILC-3 (figure 4.6) (Goto et al., 2014). As a result, the availability of fucose to the gut microorganisms, and consequently bacterial population size, can be regulated by the immunological status of the animal. This is illustrated by experiments using mice administered LPS (lipopolysaccharide: a component of the cell wall of Gram-negative bacteria, and a potent elicitor of the mammalian immune system). Because LPS does not contribute to the virulence of a pathogen, challenging an animal with LPS enables the researcher to discriminate between immune responses and direct deleterious effects of the pathogen on the host. When LPS is injected intravenously into mice, gut glycoproteins are fucosylated via a further cytokine, IL-23, released from dendritic cells (figure 4.6). The interpretation that this immune-mediated regulation of carbohydrate metabolism contributes to sustained host-symbiont relations, even when the animal is infected by a pathogen, is confirmed by parallel experiments using mice administered with the bacterial pathogen *Citrobacter rodentium*. In these mice, the enhanced fucosylation promoted the populations of *B. theta* and reduced pathogen-induced damage to the gut epithelium (Pickard et al., 2014).

4.3. The Effects of the Microbiota on Animal Immune Function

4.3.1. DIRECT EFFECTS

Just as the animal immune system plays a key role in determining the microbiota in the animal body (section 4.2), microorganisms can alter the immunological function of the animal host. In other words, the arrow of causation has to be drawn in both directions between the immune system and microbial communities. The consequences can be complex and variable: microorganisms can promote or dampen immune system function (or specific

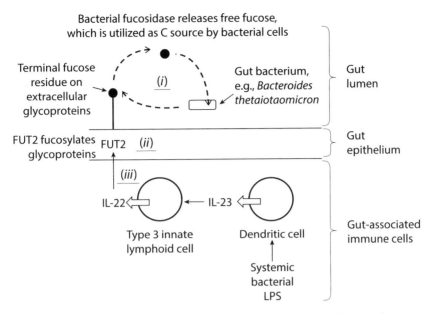

FIG. 4.6. Immune system–mediated provisioning of a key carbon source, the sugar fucose, to the gut bacterium *Bacteroides thetaiotaomicron*. (i) The fucosidase enzyme of *B. theta* releases fucose from host glycoproteins, providing free fucose for the carbon and energy requirements of the bacterium. (ii) The animal enzyme FUT2 (fucosyl-transferase) mediates the insertion of a terminal fucose residue onto mucins and other glycoproteins. (iii) The expression of *FUT2* gene is induced by interleukin-22 (IL-22) produced by type 3 innate lymphoid cells, regulated by unknown signals of bacterial origin (not shown) and IL-23 derived from dendritic cells. (Drawn from data in Goto et al. [2014] and Pickard et al. [2014].)

elements of the immune system), and these effects of the microbiota can be advantageous to one or both of the microbial and animal partners.

Many of the effects of microorganisms on the animal immune system involve molecular interactions between microbial products and the signaling cascades that either regulate the production of humoral immune effectors or alter the viability and function of immune cells. Most of the literature on these direct effects concerns the manipulation of the animal immune system by pathogens, with evidence that pathogen-induced immunological changes promote colonization, proliferation, or transmission of the pathogen, and often contribute to disease symptoms. There is, however, increasing evidence that certain resident microorganisms and their products can also modify specific immune processes. These direct effects should be distinguished from indirect consequences of microorganisms, including where immune function is altered by microbial effects on nutrition. For example, animals may become immunocompromised by infection with pathogens that drain host nutritional resources, or by perturbation of microorganisms that provide nutrients.

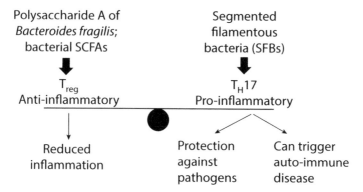

FIG. 4.7. Relationship between resident microorganisms and immunological status of the host. The immunological balance between pro- and anti-inflammatory T cells is modulated by gut microorganisms (see text for details).

A notorious instance of the direct effects of nonpathogenic microorganisms on host immune function emerged from the discovery that the immunological traits of a single inbred strain of mouse, C57BL/6, differed between two suppliers: the titer of T_H17 cells (a subset of CD4+ T helper cells) in the distal ileum of the gut was considerably greater in mice from Taconic Farm than from the Jackson Laboratory. A role for the gut microbiota was indicated by the observation that germ-free mice (which have no T_H17 cells in the distal ileum) gained T_H17 cells when colonized with the gut microbiota of Taconic mice, but not Jackson mice. Ivanov et al. (2009) identified the causal microorganisms: segmented filamentous bacteria (SFBs, Gram-positive bacteria allied with Clostridia), which were found adhering to the surface of the gut epithelium of the distal ileum of Taconic mice, but not Jackson mice. The immunological consequences of the SFB-induced T_H17 cells are very substantial. The T_H17 cells are proinflammatory, producing several cytokines, including IL-17 and IL-22. This heightened immune-responsiveness can confer protection against bacterial pathogens (Ivanov et al., 2009), but misregulation of T_H17 cells has been implicated in autoimmune diseases, including rheumatoid arthritis and possibly multiple sclerosis (figure 4.7, right).

The proinflammatory effects of T_H17 cells in the gut are counterbalanced by another set of T helper cells called regulatory T cells (Tregs) which are generally anti-inflammatory and promote immunotolerance (figure 4.7, left). Importantly, the Treg populations are strongly influenced by members of the gut microbiota. In particular, one subset of Tregs that produce the anti-inflammatory cytokine IL-10 is induced by polysaccharide A, a component of the cell wall of a common gut bacterium *Bacteroides fragilis* (Round and Mazmanian, 2010), and more generally by short chain fatty acids (SCFAs), which are waste products of anaerobic respiration of various bacteria including many clostridia and bifidobacteria (P.M. Smith et al., 2013). In this way, the products of bacteria that promote Tregs can protect mice against colitis caused by overactivity and inflammation of the GI tract.

Although most research on the effects of the microbiota on immune responsiveness of animals has been conducted on mammals, especially the laboratory mouse, there is parallel evidence that resident microorganisms can influence the immunological response of various invertebrates (figure 4.8). Notably, hemocyte numbers are depressed in pea aphids colonized by the secondary symbiotic bacteria *Hamiltonella* and *Regiella* (figure 4.8A) (Schmitz et al., 2012), and symbiont modulation of the cellular immune response is also displayed by *Vibrio fischeri*, the luminescent bacteria in the light organ of the squid *Euprymna scolopes*. Intriguingly, the many phagocytic hemocytes that reside in the light organ bind avidly and phagocytose most bacteria, but display little or no response to the native light organ symbiont, *V. fischeri*. By contrast, hemocytes isolated from squid treated with antibiotic to eliminate the *V. fischeri* are very active against *V. fischeri*, suggesting that the *V. fischeri* cells suppress hemocyte activity against themselves (figure 4.8B) (Nyholm et al., 2009). However, the interactions between the microbial symbionts and the immune system in these invertebrate animals are not yet understood at the molecular level.

4.3.2. EFFECTS AT A DISTANCE

So far, our focus on the interactions between the microbiota and animal immune system has concerned the specific organs or tissues with which the microorganisms are associated, including how locally produced immune effectors restrict microbiota abundance and composition, and how the microbiota affect local immune function. There is a further issue: the extent to which the microbiota affects the overall immune function of the animal.

Various mechanisms function to dampen the systemic consequences of an animal's immune response to the gut microbiota. As considered in section 4.2.1, the IgA+-B cells reactive against microorganisms in the mammalian gut do not circulate freely throughout the lymphatic or blood system, but are restricted to the mesenteric lymphatic system associated with the gut. Other mechanisms suppress the transfer of microbial products that activate the innate immune system from the gut to the internal body fluids. An amidase (PGRP-LB) that degrades bacterial cell wall peptidoglycan (PGN) is secreted into the gut lumen of the *Drosophila* gut, functioning to eliminate immunologically active PGN fragments that would otherwise pass from the gut lumen to the fat body, where they would trigger a systemic immune response (Zaidman-Remy et al., 2006).

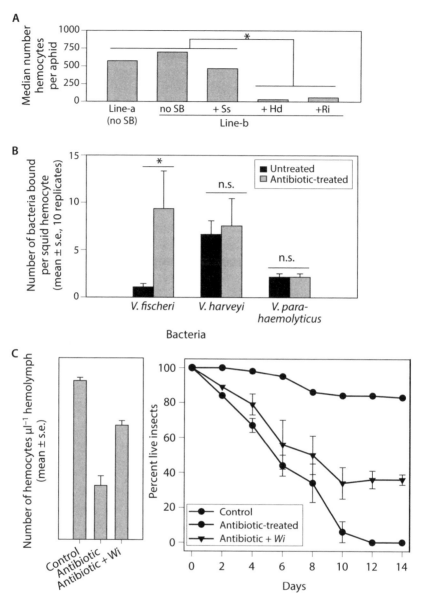

FIG. 4.8. Impact of resident microorganisms on cellular immunity. A. Number of hemocytes in two lines of the pea aphid *Acyrthosiphon pisum* (line-a, which naturally lacks secondary bacteria (SB) and line-b experimentally manipulated to bear no secondary bacteria, *Serratia symbiotica* (*Ss*), *Hamiltonella defensa* (*Hd*), or *Regiella insecticola* (*Ri*). (Redrawn from Fig. 5A of Schmitz et al. [2012].) B. Capacity of hemocytes isolated from untreated bobtail squid *Euprymna scolopes* to bind to the native symbiont (*V. fischeri*) is suppressed, relative to squid treated with antibiotic for 5 days to eliminate the symbiotic bacteria, but hemocyte binding to non-symbiotic *Vibrio* species is not significantly affected by the antibiotic treatment. (Redrawn from Fig. 3 of Nyholm et al. [2009].) C. (left) Number of hemocytes in the tsetse fly *Glossina morsitans* bearing the symbiotic

The systemic immune system is not, however, totally isolated from the interactions between the gut and its microbiota. This is revealed particularly clearly by research on germ-free mice or with a microbiota much depleted by antibiotic treatment. These mice have multiple immunological perturbations. In particular, when bacterial pathogens, such as *Listeria monocytogenes* or *Staphylococcus aureus*, are injected intravenously (to avoid any direct contact with the gut microbiota), most germ-free mice are killed, but mice with a gut microbiota recover. The susceptibility of germ-free mice to systemic bacterial infections has been attributed to deficiencies in the development and differentiation of innate immune cells, which are derived from common myeloid progenitor cells in the bone marrow (Khosravi et al., 2014). Although the production of the common myeloid progenitor cells is normal in germ-free mice, the subsequent differentiation of these cells is perturbed, resulting in depressed numbers of key innate immune cells—macrophages, monocytes, and neutrophils—available to provide protection in the first hours of the bacterial infection. Consistent with the role of the gut microbiota in promoting differentiation of myeloid lineage cells, conventional mice treated with antibiotics that deplete the gut microbiota have reduced numbers of differentiated myeloid cells, and germ-free mice colonized with live (but not killed) microorganisms from the cecum of conventional mice recover myeloid cell populations typical of conventional mice.

Complementary evidence that the gut microbiota can have long-distance effects on immune function comes from research on autoimmune diseases. For mouse models of certain autoimmune diseases, the presence and composition of the gut microbiota are a strong predictor of the incidence of the disease phenotype. In the mouse model of rheumatoid arthritis (inflammation of the joints) and experimental encephalomyelitis (a model of multiple sclerosis, caused by inflammatory demyelination of neurons of the CNS), germ-free or antibiotic-treated mice display greatly attenuated or no disease symptoms (Wu and Wu, 2012). The underlying mechanisms are not fully resolved, but likely involve reduced T_H17 cell-mediated inflammation in the absence of the gut microbiota. The gut microbiota may, however, protect against some autoimmune diseases. In particular, an important tool for biomedical research on type 1 diabetes (T1D), in which T cells destroy the insulin-producing β-cells of the pancreas, is provided by nonobese diabetic (NOD) mice. The disease phenotype is very reliable in germ-free mice but frustratingly variable in conventional mice, and the cause of this difference may be specific members

bacterium *Wigglesworthia* (control), treated with antibiotic to remove the *Wigglesworthia*, and with *Wigglesworthia* extract administered to antibiotic-treated flies (antibiotic + Wi) (Redrawn from Fig. 4A of Weiss et al. [2012]); and (right) survival of tsetse flies following challenge with *E. coli* injected into the fly hemolymph. (Redrawn from Fig 5A of Weiss et al. [2012].)

of the microbiota present in some, but not all, mouse stocks that protect against the autoimmune phenotype. Intriguingly, SFBs (which generally enhance proinflammatory T_H17 cells, see figure 4.7) have been implicated in suppression of T1D in male NOD mice but not females (Kriegel et al., 2011), suggesting that, in certain sex-specific contexts, T_H17 cells may dampen immune responsiveness. These studies raise the possibility that the composition or functional traits of the gut microbiota may influence the susceptibility of humans to certain autoimmune diseases (Bhargava and Mowry, 2014). Consistent with this interpretation, differences in the composition of the microbiota between healthy people and patients with certain autoimmune diseases have been reported; for example, people with rheumatoid arthritis commonly have high populations of the bacterium *Prevotella copri*, which induces inflammatory response when introduced to mice (Scher et al., 2013). However, more research is needed to establish the causal basis of these correlations (see chapter 3, section 3.3.2 where correlation and causation are discussed). In particular, definitive evidence whether and how specific taxa and microbial communities contribute to the disease is largely lacking, and some differences between the microbiota of healthy and unhealthy people may result from the generalized malaise caused by the disease.

Further data implicate the gut microbiota in allergic responses, including asthma, hay fever, rhinitis, and food allergies. Allergic inflammation of the airways is mediated by high serum levels of IgE and elevated basophil populations in the peripheral blood system, both of which can be linked to a disturbance in the immune balance in the gut. Consistent with a role of the gut microbiota, germ-free mice display abnormally high IgE levels, together with extreme sensitivity to oral allergens that are functionally equivalent to food or pollen allergies in humans. The role of the gut microbiota in modulating allergic inflammation is complex, and includes promotion of T_H1 cells over T_H2 cells (which mediate allergic inflammation), and reduced invariant natural killer T (iNKT) cell populations in the gut (Olszak et al., 2012). The effect of the microbiota on the balance between T_H1 and T_H2 cell populations is abrogated in mice for which a key innate immune receptor for microbial products (TLR4) is genetically ablated (Bashir et al., 2004), but the specific microbial ligands that protect against allergic inflammation have not been identified definitively. Intriguingly, this immunological disorder is alleviated by colonization of the mice with the diverse community of microorganisms associated with conventional mice, but not by any single microbial taxon tested nor by low diversity sets of microorganisms that are representative of the range of functions in the conventional mouse (e.g., the Schaedler microflora). These data are consistent with the hypothesis that the sharp rise in human atopic disease in recent decades may be linked to a reduced diversity of microbial colonists of the gut, linked to excessively hygienic conditions and possibly antibiotic treatments (see chapter 3, section 3.5).

4.3.3. MICROORGANISMS AND MATURATION OF THE IMMUNE SYSTEM

As discussed above, germ-free mice and antibiotic-treated mice are important tools contributing to our understanding of the interactions between the animal immune system and resident microbiota. Although these two treatments yield similar immunological responses in many experimental designs, the germ-free mice have never interacted with microorganisms during their life (Smith et al., 2007), and, as discussed in section 4.3.2, some aspects of their immune system are not fully developed.

Why are the final steps in maturation of the immune system in mice, and presumably mammals generally, dependent on colonization by microorganisms? It can be argued that this is highly adaptive under natural conditions. Microbial products are a superbly reliable cue for the developmental transition from the usually sterile conditions in utero, when a functional immune system is not required, to the microbe-infested world inhabited by the postnatal animal, where the immune system is essential for survival.

The developmental strategy of using microbial products as a cue for finalizing maturation of the immune system has one key weakness, that it offers a window of opportunity for pathogens to exploit the neonatal animal before its immune system has matured. The mammalian reproductive system includes multiple adaptations that reduce the hazard of infection of the newborn. These adaptations include exposure of the young to protective vaginal microbiota of the mother during parturition (see chapter 3, section 3.3.1) and promotion of protective taxa that can utilize the complex oligosaccharides in the first milk food (chapter 6, section 6.4.3), as well as maternal antibodies and other immune effectors in the milk (Sela and Mills, 2010). This strong maternal control over the first microbial colonists of the newborn is universal among mammals, with the partial exception of human societies where Cesarean delivery, bottle-feeding, and excessive cleanliness are common practice.

Is the role of microbes in the maturation of the immune system unique to mammals, or more widely distributed across the animal kingdom? This question has rarely been addressed, but intriguing data come from research on the tsetse fly *Glossina* which is obligately associated with a vertically transmitted bacterium *Wigglesworthia* (Weiss et al., 2012). When the bacteria are excluded by maternal antibiotic treatment, the bacteria-free offspring develop to adult insects with an immature and perturbed immune system. In particular, they have much depleted populations of phagocytic hemocytes, they display aberrant expression of antimicrobial peptides, and they are killed by an *E. coli* strain that is not pathogenic to unmanipulated flies (figure 4.8C). Remarkably, *Wigglesworthia*-free flies injected with an extract of the *Wigglesworthia* bacteria displayed a much-improved immunological program, including a significantly greater capacity to produce hemocytes, together with several humoral immune effectors (phenoloxidase and DUOX) and resistance to *E. coli*. The identity of the *Wigglesworthia* elicitor that mediates immune function remains to be identified.

4.4. Symbiont-Mediated Protection: Microbiota as the Second Immune System

4.4.1. PROTECTIVE MECHANISMS

In section 4.3, we have considered how microorganisms can modulate the immune function of animals. In addition to these effects, there is abundant evidence that the microbiota also contributes directly to the defense against pathogens and parasites in many animals. For example, *Hydra* are readily infected by the fungus *Fusarium* when the bacterial communities associated with their epithelial cells are depleted (Fraune et al., 2015); marine isopods of the genus *Santia* gain protection from a surface layer of *Synechococcus*-type cyanobacteria (Lindquist et al., 2005); and, compared to untreated mice, antibiotic-treated mice with a depleted gut microbiota are susceptible to up to 1,000 times lower dose of *Salmonella* (Miller et al., 1957). The selective advantage to the animal host can be profound. A North American *Drosophila* species, *D. neotestacea*, is parasitized by a nematode parasite *Howardula*, which causes reproductive sterility of the female flies, but the deleterious effect of the parasite is strongly ameliorated in flies that bear a maternally inherited *Spiroplasma* symbiont (figure 4.9A). *Spiroplasma* confers a very substantial selective advantage in natural populations of *D. neotestacea*, resulting in the spread of *Spiroplasma*-colonized flies from east to west across the continent (Jaenike et al., 2010).

The modes of action of symbiont-mediated protection are usefully described in ecological terms (Gerardo and Parker, 2014) (figure 4.10). In interference competition, the organisms interact directly, for example by producing a toxin; in exploitative competition, the competing organisms deplete the resource without direct interaction; and in apparent competition, the resident microbiota induces a host immune response that is more deleterious to their competitor (a pathogen or parasite) than to themselves. Apparent competition can be considered as an extension of the microbial modulation of host immune function discussed in section 4.3.

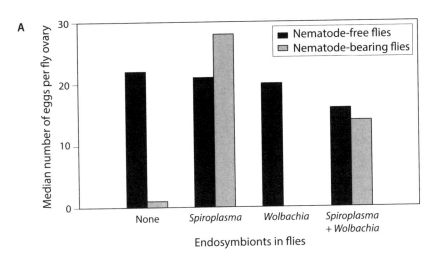

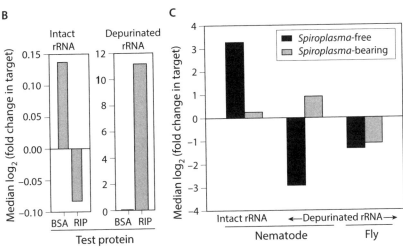

FIG. 4.9. *Spiroplasma*-mediated protection of *Drosophila neotestacea* against the nematode *Howardula aoronymphium*. A. Fertility of *D. neotestacea* bearing different endosymbionts and either infected or uninfected with *Howardula* nematodes. (Redrawn from Fig. 1A of Jaenike et al. [2010].) B. RT-qPCR based index of depurination for *Howardula* 28S rRNA incubated with recombinant *Spiroplasma* RIP protein for 4 h at 21°C, with bovine serum albumin (BSA) as control protein. (Redrawn from Fig. 4 of Hamilton et al. [2016].) C. RT-qPCR based index of abundance of intact and depurinated 28S rRNA in *Howardula*-infected *Drosophila* that are either *Spiroplasma*-free or colonized with *Spiroplasma*. (Redrawn from Fig. 5 of Hamilton et al. [2016].)

4.4.2. INTERFERENCE COMPETITION

The benefits that animals derive from associating with microorganisms that synthesize protective toxins reflects the very broad and diverse metabolic capabilities of many microorganisms (chapter 1, including Table 1.1A). Research on protective microbial toxins has been fueled by the recognition that microorganisms associated with animals represent a valuable source of novel bioactive compounds with pharmaceutical value, e.g., as antimicrobials or for cancer therapy, and by the increasing availability of genomic data enabling predictions of secondary metabolite biosynthesis capabilities in unculturable microorganisms. The best-known examples of these defensive symbioses are in sessile marine animals, such as sponges, bryozoans, and ascidians, and in various insects, but there is every expectation that microbes with defensive functions are very widespread across the animal kingdom. The taxonomic range of microorganisms and chemical diversity of their products are, similarly, very broad (Florez et al., 2015).

Some animals derive protection from a single compound produced by a single microbial partner. For example, the *Spiroplasma* that protect *Drosophila neotestacea* from *Howardula* nematodes produce a ribosome-inactivating protein (RIP) that depurinates RNA, damaging the rRNA and consequently the ribosome function of the *Howardula* (figure 4.9B); this effect is specific to

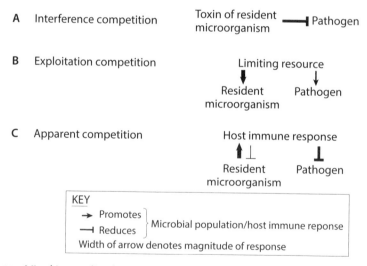

FIG. 4.10. Microbiota-mediated protection of animals against pathogens or other natural enemies. A. Interference competition: direct interaction between resident microorganism(s) and pathogen, reducing capacity of pathogen to colonize and proliferate in the host. B. Exploitation competition: resident microorganism(s) and pathogen utilize a common resource that is limiting, reducing pathogen proliferation and persistence. C. Apparent competition: resident microorganism(s) induces a host immune response which is more deleterious to the pathogen than to itself.

the *Howardula*, without affecting the rRNA of the fly (figure 4.9C) (Hamilton et al., 2016). Other apparent examples of protection by single compounds include the embryos of the shrimp *Palaemon macrodactylus* that are invariably colonized by bacteria of the genus *Alteromonas*, which produce the secondary metabolite 2,3-indolinedione, conferring protection against a major fungal pathogen, *Lagenidium callinectes* (Gil-Turnes et al., 1989) and the pseudo-monad bacterium in *Paederus* rove beetles, which codes for the synthesis of a single onnamide-like polyketide, informally known as pederin (Piel, 2002).

In many animals, however, the microorganisms contributing to animal defense produce a cocktail of protective compounds, a trait that is predicted to hinder the evolution of resistance in natural enemies. In colonial ascidians, this trait is attributed to the possession of multiple symbionts with different metabolic capabilities. These animals have an intracellular α-proteobacterial symbiont *Candidatus* Endolissoclinum faulkneri that produces polyketides (Kwan et al., 2012) and a photosynthetic cyanobacterial symbiont *Prochloron* that produces various cyclic peptides, including the patellamides (Donia et al., 2006). The hypervariable gene *patE* confers patellamide toxicity, and each ascidian host is protected by a wide diversity of patellamides because, although each *Prochloron* genome has a single *patE* gene, every host bears multiple, closely related *Prochloron* genotypes with different *patE* genes.

In some associations, a wide chemical diversity of protective toxins is not achieved by the accommodation of multiple microbial genotypes, but by associating with a single microorganism that can synthesize many secondary compounds. Particularly remarkable secondary chemistry is displayed by filamentous bacteria of the genus *Entotheonella* in marine sponges. A single phylotype of *Entotheonella* TSY1 in the sponge *Theonella swinhoei* has multiple gene clusters for the synthesis of onnamide polyketides, polytheonamide peptides, various non-ribosomal peptides, and additional secondary metabolites that have yet to be identified. The predicted metabolites are readily detectable in the sponge tissue by mass spectrometry (Wilson et al., 2014). As a result of their exceptional complement of secondary metabolite biosynthesis genes, the *Entotheonella* symbionts have among the largest known bacterial genomes, at 9–10 Mb. The selection pressures for the evolution of these multifunctional "do-it-all" symbionts is encapsulated in the Black Queen hypothesis (Morris et al., 2012). Specifically, functions that support host protection are public goods for the microbial community, i.e., the loss of the capability is selectively advantageous to the individual microorganism (which can devote more resources to growth and proliferation), while conferring minimal selective disadvantage at the level of the entire association. A community is predicted to comprise multiple taxa or genotypes that do not contribute to the public good, and one or a few taxa/genotypes that (like the Black Queen in the card game) mediate these crucial functions. In the context of the community, the Black Queen genotype is under exceptionally strong selection pressure to retain the functions on which the survival of the association (and therefore itself) depends.

At first sight, it is perhaps surprising that toxin-producing symbionts are very widespread across the animal kingdom. These associations are likely to be costly to the host because the microorganisms require resources and space under conditions where protection is not required, and the microbial products may be toxic to the host or other beneficial microbes. The maintenance cost of a protective symbiont is neatly illustrated by the relationship between the pea aphid *Acyrthosiphon pisum* and the bacterium *Hamiltonella defensa*, which confers resistance to the parasitic wasp *Aphidius ervi*. When pea aphids bearing and lacking *Hamiltonella* were maintained together over multiple generations, the frequency of aphids with *Hamiltonella* rose to >95% in the presence of the parasitoid, but declined significantly in the absence of the parasitoid (Oliver et al., 2008). The costs of *Hamiltonella* in the absence of the natural enemy, together with evidence that many pea aphid genotypes have intrinsic resistance to the parasitoid, may account for the intermediate frequency of this bacterium in natural pea aphid populations.

Other protective symbionts that are only intermittently required are, however, retained in all host individuals through the life cycle. For example, survival of the pupal stage of the solitary wasps of the tribe Philanthini (bee-wolves) is absolutely dependent on antibiotic-producing *Candidatus* Streptomyces philanthi, which prevents infection by pathogenic fungi (Kroiss et al., 2010). Other life stages of the insect maintain the association without requiring this function. The *Streptomyces* are stored in glandular cavities in the adult antennae and inoculated onto the ceiling of the brood chamber by antennal smearing; the larvae then incorporate the bacteria into the cocoon that they spin around their body, prior to pupation (Kaltenpoth et al., 2005). This remarkable suite of insect morphological and behavioral adaptations, together with phylogenetic evidence for strong partner fidelity between the insect and *Streptomyces* lineages for ca. 70 million years (Kaltenpoth et al., 2014), can be attributed to the critical importance of the protective symbiosis during the immobile and vulnerable pupal life stage.

4.4.3. EXPLOITATION COMPETITION

Exploitation competition refers to competition for limiting resources (figure 4.8). The contribution of exploitation competition to protection against pathogens is exemplified by an elegant study on interactions between the gut microbiota and the pathogen *Citrobacter rodentium* in the mouse gut (Kamada et al., 2012). *C. rodentium* is a natural pathogen of mice and widely used as a model for pathogenic *E. coli* (enteropathogenic and enterohemorrhagic;

EHEC and EPEC) because these different bacteria interact with the intestinal epithelium in the same way. *C. rodentium* is more virulent to germ-free mice than specific pathogen–free (SPF) mice. The SPF mice can contain the bacterial infection to 10-fold lower levels than germ-free mice, and can clear the infection over three weeks, while germ-free mice remain infected for at least 6 weeks. The immune response of the SPF and germ-free mice to the *C. rodentium* does not differ significantly, and the role of the microbiota is suggested by clearance of the infection in germ-free mice when they are colonized with gut microbes, either by oral administration or by cohousing with SPF mice. Further experiments demonstrated that the gut microbiota confer protection against *C. rodentium* by competing for simple sugars required by the pathogen for growth.

Although many instances of exploitation competition involve two microbial taxa that compete for a specific resource limiting their growth and proliferation, other instances of exploitation competition are mechanistically complex. This is beautifully illustrated by research on interactions between the intracellular bacterium *Wolbachia* and viruses. Although *Wolbachia* is best known as a reproductive parasite of insects and other arthropods, it can also act as a protective symbiont, particularly against viruses. For *Drosophila melanogaster*, in which this effect was first demonstrated (Hedges et al., 2008; Teixeira et al., 2008), various viruses cause much greater mortality in *Wolbachia*-free *Drosophila* than in *Drosophila* bearing *Wolbachia*, although the magnitude of the effect varies with virus and *Wolbachia* genotype. Research on the interactions with the *Drosophila* C virus (DCV) reveals that *Drosophila* is protected, at least in part, by competition for a key limiting resource: cholesterol (Caragata et al., 2013). Cholesterol is a sterol and, because *Drosophila* cannot synthesize sterols, it obtains its sterol requirements from the diet. However, when the standard *Drosophila* diet is supplemented with additional cholesterol, the cholesterol content of the flies increases, and the protective effect of *Wolbachia* against DCV is abrogated, i.e., the DCV attain higher titers and the flies die more quickly (figure 4.11). Control experiments confirmed that dietary cholesterol does not affect either the *Wolbachia* titer in the flies, nor the survival of *Wolbachia*-free flies. Cholesterol is a vital constituent of the DCV envelope because, although dengue virions (DENV) depleted of cholesterol are endocytosed into host cells, they cannot escape from the endosome into the host cell cytoplasm, and so cannot replicate (Carro and Damonte, 2013). It has been argued that the *Wolbachia* population may deplete the cholesterol available in the *Drosophila* for the DENV envelope, not because it has any requirement for cholesterol but because each *Wolbachia* is bounded by an individual host membrane, resulting in the large-scale sequestration of cholesterol to the *Wolbachia*-infected cells.

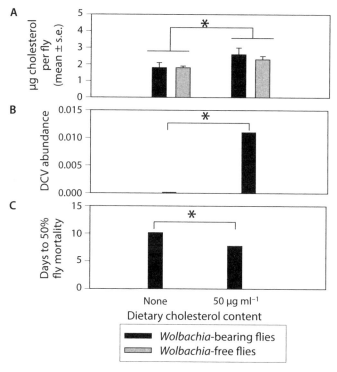

FIG. 4.11. Exploitation competition between *Wolbachia* and *Drosophila* C virus (DCV). A. Dietary cholesterol significantly increases the cholesterol content of *Drosophila*. (Redrawn from Fig. 1E of Caragata et al. [2013].) B. The DCV titer in *Wolbachia*-colonized flies is increased on the high cholesterol diet. The abundance of DCV was quantified as the ratio of DCV:*cycK* gene of *Drosophila*. (Redrawn from Fig. 2B of Caragata et al. [2013].) C. Dietary cholesterol reduces the survival of flies colonized with *Wolbachia* (strain *w*MelCS) challenged with DCV. (Redrawn from Table 1 of Caragata et al. [2013].)

4.4.4. APPARENT COMPETITION

Apparent competition mediated by the host immune system has been demonstrated in various interactions involving pathogens. For example, the upper respiratory tract of mice is readily colonized by single infections of the bacteria *Streptococcus pneumoniae* and *Haemophilus influenzae*, but coinfection with both bacteria results in the rapid decline and loss of *S. pneumoniae*. This interaction is not caused by direct competitive interactions between the two bacteria but by the recruitment and activation by *H. influenzae* of host neutrophils that selectively kill *S. pneumoniae* cells (Lysenko et al., 2005). The host immune system also mediates apparent competition between the gut pathogen *Salmonella enterica* serovar Typhimurium and the resident microbiota. The *Salmonella* virulence factors induce a strong inflammatory response in the gut epithelium, which selectively eliminates most of the resident microbiota (Stecher et al., 2007). The host immune response is also directly advantageous to the *Salmonella* because reactive oxygen species generated in the inflammatory response react with thiosulfate in the gut lumen to generate tetrathionate, which is used by *Salmonella* as a respiratory electron receptor (Winter et al., 2010).

Does apparent competition also contribute to microbiota-mediated protection against pathogens? Definitive data are largely lacking, but the gut microbiota has been implicated in promoting antiviral response in *Drosophila* (Sansone et al., 2015). Specifically, antiviral effectors generated in gut epithelial cells are induced by ERK signaling cascade in response to the cytokine Pvf2 (Pvf proteins of *Drosophila* are akin to the PDGF/VEGF factor of vertebrates). However, Pvf2 production is under dual control requiring both signaling from viral factors via Cdk9 and activation of the NF-κB transcription factor by bacteria in the gut (figure 4.12). In this way, the bacteria activate an immune response that is deleterious to orally acquired viruses with minimal negative effect on themselves.

4.5. Summary

The immune function of an animal is the product of multiple and diverse interactions with its microbiota. In the healthy animal, the response of the immune system to the microbiota and the reciprocal effects of the microbiota on the immune system generate a community of benign and beneficial microorganisms interacting with an immune system of appropriate reactivity. Perturbation of this balance by dysfunction of either the immune system or the microbiome results in ill-health. Because of the multiplicity of interactions, the rules of engagement between the microbiota and host immune system are predicted to be highly dynamic. Nevertheless, some general themes are emerging.

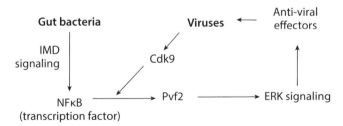

FIG. 4.12. Apparent competition between gut bacteria and viruses. The gut bacteria activate antiviral ERK signaling in the gut epithelial cells of *Drosophila* via an immune transcription factor NFκB, which Induces the expression of the *pvf2* gene under conditions of virus-stimulated signaling via Cdk9. (From data in Sansone et al. [2015].)

One key theme is the central role of microbicidal immune effectors in regulating the abundance and distribution of the microbiota (section 4.2). In the mammalian gut, products of the adaptive immune system, especially IgAs, and the innate immune system, including lectins and antimicrobial peptides, collaborate to limit most gut microorganisms to the outer regions of the mucus, thereby avoiding direct contact between microorganisms and the gut epithelial cells. Innate immune effectors in invertebrates, similarly, restrict the abundance and distribution of microbial colonists. Differential susceptibility to these immune effectors can have a defining effect on the composition of the microbiota, with evidence that high host immunoreactivity can promote beneficial microorganisms in some systems (e.g., *Vibrio* in squid) but have the reverse effect in other systems (e.g., the gut microbiota of *Drosophila*).

A second theme is that the microbiota regulates the immune function of the host by multiple different interactions, some of which promote immune reactivity and others that dampen immune responsiveness (section 4.3). These effects are particularly evident in relation to the cellular immune system, including the balance of pro- and anti-inflammatory T cells of the adaptive immune system in mammals, and the abundance and properties of hemocytes in some invertebrates. These effects extend to microbial regulation of the development of the immune system, with evidence that stimulation by microbial products is required for the final stages in the differentiation of innate immune cells of mammals and possibly also of insects.

The final theme is that the animal immune system is a subset of the global defensive system of animal. This is because resident microorganisms play a crucial protective role (section 4.4). In some cases, the protective effects are mediated by resource capture (exploitation competition), i.e., the resident microorganisms deplete the availability of limiting resources required by the incoming pathogen, or by stimulation of the host immune effectors that are selectively deleterious to the incoming pathogen (apparent competition). In addition, various microorganisms produce compounds that are toxic to natural enemies of their animal host. This effect, which is known as interference competition, can involve specific microbial partners with the genetic capacity to produce one or many secondary metabolites, or communities of microorganisms with diverse metabolic capabilities. How these microbial-mediated defensive capabilities interact with the intrinsic immune system of the animal is largely unknown. However, as will be discussed in chapter 7 (section 7.2.3), there is initial evidence that acquisition of microbial defense can lead to evolutionary changes in the responsiveness of the animal immune system.

5

Microbial Drivers of Animal Behavior

5.1. Introduction

Decades of research on pathogens have revealed that microorganisms can drive the behavior of animals (figure 5.1A). In some instances, the pathogen-induced behavior promotes the fitness of the pathogen and not the host: this is microbial manipulation of host behavior. For example, grasshoppers infected with the fungal pathogen *Entomophaga grylli* display aberrant behavior, known as summit disease. They climb to the top of vegetation, where they die, facilitating the aerial dispersal of fungal spores released from the cadaver. Similarly, rats infected with *Toxoplasma gondii* have no fear of cats and may even be attracted to the odor of cat urine, increasing the opportunity for predation and transfer of the pathogen to the cat, which is the primary host of the *Toxoplasma* (House et al., 2011). Other pathogen-induced behavioral changes are adaptive for the host, and can be described as behavioral defense (Hart, 1994). A very wide range of pathogens induce adaptive sickness behavior in humans and other mammals, including reduced appetite and locomotory activity accompanied by fever and sleep, collectively promoting clearance of the pathogen and repair of the pathogen-induced damage (Dantzer et al., 2008). In an analogous way, many insects respond to fungal infection by changing their behavior to seek out a warmer environment and the resultant increase in body temperature, which is known as behavioral fever, can eliminate temperature-sensitive pathogens with demonstrated fitness advantage to the insect (Elliot et al., 2002). Although host behavioral defenses have traditionally been treated as independent of microbial manipulation of host behavior, these two processes interact: an infected animal can exhibit some behavioral traits that are defensive and others that are driven by the pathogen, and behavioral traits can change over evolutionary time, driven by opposing selective interests of host and pathogen (Ezenwa et al., 2016).

The wealth of evidence that the behavior of animals both affects pathogen fitness and is affected by pathogens provides the impetus to consider whether animal behavior may also interact reciprocally with resident microorganisms. In some respects, the dynamics of the interactions between resident microorganisms and animal behavior are predicted to match the dynamics of host-pathogen interactions (figure 5.1A), although the effects may be more subtle than in host-pathogen interactions. The microbial partners are under selection to manipulate host behavior to their own advantage, while host behavioral traits (grooming, food selection, etc.) may control the abundance and composition of associated microbiota. However, there is some predicted overlap in selective interest between host and resident microorganisms, and

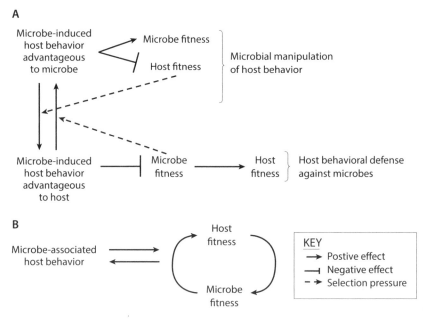

FIG. 5.1. Interactions between microorganisms and animal behavior. A. Antagonistic interactions: microorganisms induce changes in host behavior that are beneficial to either the microorganism (top) or host (bottom), but not both partners, and the opposing selection pressure on microorganism and host can driver counter-adaptations in the other partner (these counter-adaptations can alter host behavioral traits, as displayed, or other traits, e.g., immunity, metabolism). B. Mutualistic interactions: microbe-associated host behavior promotes the fitness of both the host and microorganism, e.g., host behavioral adaptations that increase microbial transmission between related hosts.

this can translate into host behavioral traits that are adaptive for both partners (figure 5.1B). For example, host behavioral adaptations favoring transmission of microbial partners to offspring or other kin promotes the inclusive fitness of the host as well as increasing the abundance and distribution, and hence fitness, of microbial partners. We should, therefore, expect both antagonistic and mutualistic interactions between the behavior of animals and traits of their resident microorganisms.

The framework for this chapter comprises two of Tinbergen's four fundamental questions in behavior research: How?—the proximate question of the mechanisms by which the presence and activities of microorganisms influence animal behavior, and Why?—the ultimate question of the selective advantage of microbial influence on animal behavior. Although the behavioral consequences of microbes in healthy animals have traditionally been neglected, in comparison to the behavior of pathogen-infected individuals, evidence for microbiota effects is now accumulating, especially in relation to three broad categories of behavior: feeding behavior, including appetite and food choice (section 5.2); the emotional state and mental well-being of mammals, including humans (section 5.3); and animal communication (section 5.4).

5.2. Microbes and Animal Feeding Behavior

5.2.1. HOW THE MICROBIOTA MAY INFLUENCE ANIMAL FEEDING

The feeding behavior of animals encompasses three broad traits: the amount of food consumed, the pattern of feeding (the size and duration of the meal, and the time interval between meals), and, where alternative foods are available, food choice. Feeding by many animals is also intimately related to the regulation of locomotory behavior because feeding is commonly preceded by, first, foraging behavior (i.e., movement in search of food) and then the inhibition of locomotion when the animal initiates feeding.

Feeding behavior has traditionally been investigated without reference to the microbiome and has often been treated, either implicitly or explicitly, as an optimality problem: how feeding behavior is regulated to obtain the energy and nutrients required to meet the demands for energy expenditure, growth, and reproduction. There is now a wealth of evidence that feeding is controlled by multiple, interacting neuronal and endocrine circuits both within the brain and between the brain and peripheral organs (Perry and Wang, 2012; Pool and Scott, 2014; Sohn et al., 2013). The gut is one of these peripheral organs, providing information about the amount and nutritional quality of the food in the gut. Other organs, especially adipose tissue, communicate the energy and nutritional status of the animal to the brain, while the sensory organs, especially the olfactory and gustatory systems, communicate information about food availability. The signaling molecules involved in the regulation of feeding are also diverse. They include multiple neurotransmitters, neuromodulators, and hormones, some of which promote appetite while others signal satiety (Perry and Wang, 2012). This complexity means that the feeding behavior of an animal can be informed by multiple aspects of its internal state.

In recent years, it has been suggested that the gut microbiota may interact with the complex regulatory network that determines animal feeding behavior. In principle, the microbiota can interact with the regulation of host feeding in two ways (figure 5.2). The first (Route-1 in figure 5.2) is by altering nutrient availability and, consequently, the dietary needs of the animal. For example, the various microorganisms that provide animals with vitamins and essential amino acids are predicted to reduce host selection of diets enriched in these essential nutrients, and microbial fermentation of complex polysaccharides into compounds (e.g., short chain fatty acids, SCFAs) that can be utilized by the animal reduces demand for dietary calories. Some microorganisms compete with the host for nutrients, e.g., many bacteria in the mammalian small intestine consume simple sugars that would otherwise be assimilated by the host (Zoetendal et al., 2012), potentially altering host behavior to increase food consumption and the choice of nutrient-rich foods. In these interactions, the effect of the microbiota on feeding behavior is defined by its effect on host nutritional status, and no microbial effects on the regulatory networks controlling feeding behavior need to be invoked.

The second way (Route-2 in figure 5.2) that microorganisms can influence feeding behavior is by interacting directly with the host signaling networks that regulate feeding behavior. By dampening or stimulating specific signaling pathways in the network, microorganisms can modify the behavioral response of an animal to its nutritional status and, thereby, influence how much the animal feeds, when it feeds, and what it feeds on.

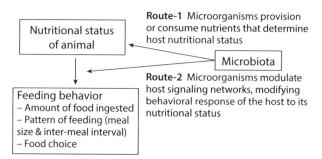

FIG. 5.2. Microbiota-mediated regulation of animal feeding behavior by effects on the nutritional status (Route-1) or signaling circuits that regulate feeding behavior (Route-2).

It is this second route that has attracted great interest. The core hypothesis is that gut microorganisms contribute to the regulation of feeding behavior, and that perturbation of the microbiota drives maladaptive feeding behavior, potentially including eating disorders in humans. Specifically, certain gut microbial communities may promote overeating and maladaptive food choices that contribute to the high incidence of obesity and chronic metabolic disease (Alcock et al., 2014).

5.2.2. THE EVIDENCE FOR MICROBIAL EFFECTS ON ANIMAL FEEDING

A direct causal link between the composition of the gut microbiota and food consumption has been demonstrated by a study on mice with a null mutation in the Toll-like receptor 5 (TLR5) gene. TLR5 plays a crucial role in innate immunity by binding the bacterial flagellar protein, flagellin, leading to activation of the adaptive immune response against flagellated bacteria and protection against bacterial invasion of the gut wall (Cullender et al., 2013). TLR5 knock-out (T5KO) mice have a perturbed gut microbiota and, for reasons that are still not fully understood, they also display hyperphagia (i.e., the mice overeat), resulting in excessive lipid deposition and hyperglycemia (Vijay-Kumar et al., 2010). Evidence that the perturbed feeding behavior is driven by the gut microbiota comes from further experiments showing that hyperphagia is abolished in antibiotic-treated knock-out mice and, second, is displayed by wild-type (WT) mice colonized with the microbiota from the T5KO mice (figure 5.3A). Further experiments on the taxonomic composition of the microbiota in the WT and T5KO mice revealed species-level differences, but it is not known how the microbial communities in the WT and T5KO mice differ in functional terms.

Several studies have implicated the fermentation products of gut microbiota in the regulation of feeding behavior. Most research has focused on one compound, acetic acid. In experiments conducted on rats (Perry et al., 2016), the supply of acetic acid to the gut lumen was controlled precisely, and, over a 10-day experiment, the rats receiving supplementary acetic acid displayed hyperphagia, accompanied by elevated plasma levels of two important proappetite hormones produced in the gut, ghrelin and gastrin (figure 5.3B). Several studies with different experimental designs have obtained similar conclusions, but other studies have yielded the reverse results, i.e., that acetic acid suppresses appetite. The reasons for this variability is not understood fully and will probably have to await a better understanding of how and where in the body the gut-derived acetic acid interacts with the regulation of feeding.

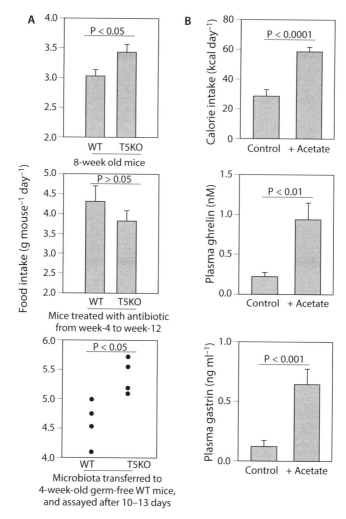

FIG. 5.3. The gut microbiota and food consumption. A. Mice with TLR5 null mutation (T5KO) display hyperphagia, relative to wild-type (WT) mice. This behavioral difference is abolished by antibiotic treatment and transferred to germ-free WT mice by colonization with microbiota from T5KO mice. (Redrawn from Fig. 3C, Fig. 4B, and Fig. 4D of Vijay-Kumar et al. [2010].) B. Hyperphagia in rats induced by intragastric infusion with acetic acid at 20 μmol kg^{-1} min^{-1}, associated with elevated plasma titers of the pro-appetite hormones ghrelin and gastrin. (Redrawn from extended data Fig. 9j and Fig. 5c & 5e of Perry et al. [2016].)

5.2.3. MECHANISMS OF MICROBIAL IMPACTS ON THE REGULATION OF ANIMAL FEEDING

As considered in section 5.2.1, the principal interest in microbial effects on feeding relate to their intervention in the animal regulatory mechanisms and signaling molecules that control food consumption and food choice. In principle, microorganisms can either synthesize some of these signaling molecules or related compounds, or influence the production of these compounds by the host. These microbial effects may be general among animals because many of the signaling molecules that regulate feeding are highly conserved across the animal kingdom. Let us examine the evidence.

Various microorganisms, including species found in animal guts, have been reported to synthesize metabolites that are bioactive in animals. For example, some bacteria release dopamine, which is a neurotransmitter that plays a central role in the reward circuits underpinning foraging behavior and feeding in many different animals (Tsavkelova et al., 2006). In addition, bioinformatic studies suggest that various bacteria have the genetic capacity to produce analogs of conserved animal signaling molecules (Fetissov et al., 2008). For example, *Bacteroides* genomes are reported to code for peptides similar to the evolutionarily related neuropeptide Y (NPY) of mammals and neuropeptide F (NPF) of various invertebrates, and *Lactobacilli* code peptides similar to the mammalian satiety hormone leptin and its insect ortholog Unpaired2. These findings are, however, far from conclusive. Definitive evidence will require studies of gut microorganisms in the gut environment, to establish whether they release bioactive compounds at physiologically relevant rates, together with parallel demonstration that host feeding behavior differs in the predicted ways between individuals bearing wild-type bacteria and mutant bacteria that cannot produce the compounds of interest.

The alternative mechanism by which microorganisms may influence feeding behavior is by modulating host production of key regulatory molecules. At this time, there is no definitive evidence for this proposed mechanism, and some candidate mechanisms proposed in the review literature lack biological realism. The activity of the neurotransmitter serotonin illustrates the complexity. Serotonin in the brain functions as a strong satiety signal, thereby influencing feeding, and most of the serotonin in the human body is produced by endocrinal cells (specifically enterochromaffin cells) of the gut, where its production is promoted by SCFAs released from the gut microbiota (Yano et al., 2015). At first sight, this reads as a clear-cut illustration of microbial effects on the regulation of feeding behavior, but this conclusion is almost certainly erroneous. The reason is that the serotonin content of the brain is metabolically independent of the serotonin synthesized in the gut. The serotonin in the gut promotes gut peristalsis, thereby regulating the rate of passage of food through the gut, and some of the gut serotonin escapes into the circulatory system, where it plays an important role in wound-healing. (Specifically, it is taken up into blood platelets, and released in response to wounding, where it promotes vasoconstriction, facilitating clot formation.) From the behavioral perspective, however, the key issue is that the blood-borne serotonin is not taken up into the adult brain. Indeed, the isolation of the small serotonin pools in the brain from fluctuating levels in the very large serotonin pools in the rest of the body is crucial for its function in the regulation of feeding behavior. This independence between titers in the central nervous system and the rest of the body, including the gut, is likely to apply to many neuroactive metabolites.

5.2.4. EVOLUTIONARY SCENARIOS

In parallel with ongoing research on the mechanisms of microbial effects on animal feeding behavior, the ultimate question "why?" has attracted considerable interest.

A useful place to start is the null hypothesis that any effect of the microbiota on host feeding is fortuitous from the microbial perspective. In other words, the selective advantage to the microorganism of microbial products that influence animal feeding behavior is unrelated to their effect on the host. Various microbial products with a candidate role in the regulation of animal feeding behavior, including products of carbohydrate degradation (e.g., acetic acid and other SCFAs) and amino acid metabolism (e.g., dopamine, tyramine), can be argued to conform to this hypothesis because these metabolic reactions serve important functions independent of the animal host. Fermentation pathways recover reducing power in the absence of aerobic respiration, powering biosynthetic reactions, and ultimately growth, of the microbial cell, and dopamine and related compounds are degradation products of amino acid catabolism. Some of these compounds can be inhibitory to other microorganisms (e.g., lactic acid, a major fermentation product released by many lactobacilli, acidifies the local environment, inhibiting bacteria that are intolerant of low pH). Other bacterial metabolites function as intraspecific signaling molecules (e.g., indole, a metabolite of the amino acid tryptophan, mediates biofilm formation and some aspects of stress resistance in various bacteria), and so they may contribute to the among-microbe interactions that shape the composition of the gut microbiota.

Even though (as indicated above) many microbial metabolites may have functions independent of their effects on animal feeding behavior, they may provide the host with accurate information about the composition and activity of the microbiota, including likely effects on host nutritional status. In other words, these microbial products may be a cue influencing animal behavior. Biological cues are discussed in chapter 2 (section 2.2.3 and figure 2.3). In the context of animal feeding behavior, microbial products can be defined as a cue where their synthesis did not evolve in relation to their information value to the host and the microbial producers of the compounds do not benefit from the behavioral response of the host.

An alternative evolutionary scenario is that the microorganisms benefit from the change in host feeding behavior triggered by the compounds they release. In this respect, the relationship between host and microbiota can be antagonistic (Alcock et al., 2014). Specifically, the abundance and fitness of certain gut microorganisms may be promoted in a host that feeds at higher rates or on different foods from the optimum for the host (figure 5.4A). The microbial-induced deviation from the host optimum may be mediated by microbial manipulation of the regulatory circuits controlling feeding, via supplementary production of neurotransmitters or hormones. Particularly intriguing is the growing evidence for dopamine-mediated plasticity of the brain in mediating acquired taste preference (Narayanan et al., 2010; Yamagata et al., 2015), providing the evolutionary opportunity for microbial-mediated perturbation of host food choice. In this respect, it has been demonstrated repeatedly that human obesity and hyperphagia are associated with reduced diversity of the gut microbiota (Zhao, 2013). It has been argued that low taxonomic diversity may be associated with functional perturbation (see chapter 3, section 3.5.1), perhaps because the dominant members of a community have a higher density and abundance in a low diversity community than a high diversity community (assuming equivalent total microbial abundance), thereby increasing their influence, e.g., total production of behavior-modifying compounds, on host traits (Alcock et al., 2014). Consequently,

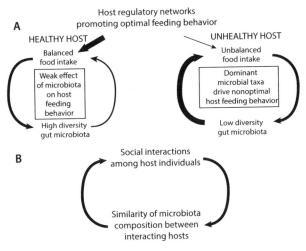

FIG. 5.4. Animal-associated microorganisms as reinforcers of behavioral traits. A. The primacy of host controls over feeding behavior (food choice and total food intake) is sustained in animals feeding on a balanced diet (left); but may be perturbed by unbalanced food intake which favors specific microbial taxa that reinforce feeding behavior advantageous to the microbiota but not the host (right). B. Social interactions between individual animals may promote similarity in the microbiota composition (through transmission of microorganisms between host contact) and be promoted by microbial effects favoring interactions between hosts with similar microbiota.

dominant members of a low diversity community can potentially manipulate host feeding behavior to favor overeating on unbalanced foods, to the selective advantage of the manipulating microbes, despite the resultant ill-health of the host (figure 5.4A).

The notion that the behavior of an animal can be trapped in a maladaptive state dictated by its microbiota has enormous appeal because, in the human context, it suggests that simple microbial therapies can lead effortlessly to healthy eating habits. What is missing is hard evidence. One line of evidence that is not supportive is that microbial manipulation of animal feeding behavior is predicted to select for evolutionary diversification of the animal neurotransmitters and hormones regulating feeding behavior, to escape deleterious microbial effects; but, contrary to this prediction, many of the neurotransmitters and hormones regulating appetite are remarkably conserved across different animal groups (see section 5.2.3). Although this pattern certainly does not refute the scenario depicted in figure 5.4A, it does bring into focus the need for rigorous testing of hypotheses in this area.

5.3. Microbial Arbiters of Mental Well-Being

5.3.1. THE MICROBIOME-GUT-BRAIN AXIS

There is now a wealth of evidence suggesting that the composition and activities of the gut microbiota can influence complex behavior of humans and other mammals, particularly emotional state. These effects of the gut microbiota are widely believed to be mediated via the gut-brain axis, meaning that the microbiota modulates the bidirectional communication between the gut and brain (figure 5.5). Virtually all the research in this area has been conducted on rodent models and humans, with a particular focus on the stress response, in which the hypothalamus-pituitary-adrenal axis (HPA) is activated to release cortisol into the blood. As figure 5.5 illustrates, one of the physiological responses to high cortisol levels is altered motility and secretory activity of the gut, and these changes in gut physiology are communicated back to the brain via neural and endocrinal pathways, leading to loss of appetite and often nausea.

It has been known for decades that the gut-brain axis plays an important role in integrating whole-animal physiology, with the gut often described as the second brain. What has become evident only recently is that the presence and composition of the gut microbiota profoundly influences—and is influenced by—the cross-talk between the brain and gut. In other words, the microbiome may be an important regulator of the gut-brain axis, and may play an important role in shaping mental health and psychiatric disorders. The best evidence comes from preclinical research, especially using the laboratory mouse model.

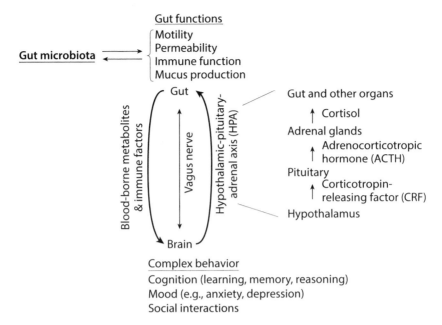

FIG. 5.5. Microbiota-mediated regulation of animal microbiota-gut-brain axis. The microbiota interacts with the bidirectional communication between brain and gut to influence the reciprocal effects on gut function and complex behavior of the animal.

5.3.2. INSIGHTS FROM RODENT MODELS

Microbial effects on the gut-brain axis are indicated by the behavioral consequences of manipulating the gut microbiota in mice. In particular, germ-free (GF) mice generally behave differently from conventional mice (i.e., mice with an unmanipulated microbiota). Valuable information about the social behavior of mice has been obtained from studies using a three-chamber assay in which the test animal is placed in a central chamber with access to two alternative chambers. In the study of Desbonnet et al. (2014), GF mice made significantly fewer social choices than conventional controls, preferring a mouse-free chamber over a chamber containing a mouse and showing no preference for a familiar mouse over a novel mouse (figure 5.6A). Furthermore, the time devoted to stereotyped grooming behavior was also extended in GF mice relative to conventional mice, and conventionalization (adding back microorganisms derived from conventional mice to the GF mice) reversed the behavior of GF mice with respect to the empty chamber experiment and grooming behavior (figure 5.6A). Other widely used assays quantify how mice respond to potentially stressful novel environments. For example, in the step-down test, a mouse is placed on a small platform and the time required for the mouse to step down and explore its surroundings is determined.

In one study, laboratory strains of mice varied in their time to step down: BALB/c mice were "timid," often taking several minutes to step down, while NIH Swiss mice were "risk-takers" that step down within seconds. Remarkably, these traits may not be driven by the mouse genotype, but by the microbiota (Bercik et al., 2011). Reciprocal transfer of the microbiota between the two mouse strains yielded NIH Swiss mice that were more timid and BALB/c mice with greater risk-taking behavior (figure 5.6B).

Despite the data from individual studies demonstrating how the presence and composition of the gut microbiota can alter animal behavior, an overview of literature (Luczynski et al., 2016) reveals considerable among-study variation in microbiota effects. Germ-free mice can display heightened or reduced levels of anxiety, and increased or reduced social interactions with other mice, and reports of increased movement and impaired memory are not fully reproducible across studies. Although much of the variation can be explained in terms of sex and strain effects, as well as the detailed design of the behavioral assays, the more important conclusion from the broad spectrum of responses is that the microbiome-gut-brain axis is very complex with many interacting factors contributing to the response to microbiota perturbations.

What are the mechanisms underlying these effects of the microbiota on behavior of mice? Multiple studies demonstrate systematic differences in brain chemistry of GF-mice relative to conventional mice, including reduced levels of brain-derived neurotrophic factors (secreted proteins that promote the survival and growth of neurons) in the cortex, hippocampus, and amygdala, and altered expression of key genes expressed at synapses (e.g., genes coding subunits of the glutamate receptor and dopamine receptor) in specific regions of the brain (Luczynski et al., 2016). The microbiota also influences the function of the HPA axis (see figure 5.5), which regulates the titer of the corticosteroid stress hormones in the blood. This was first demonstrated by experiments exposing conventional and GF-mice to a standardized environmental stress: squeezing the mouse into a 50 mL tube, where it is unable to move (Sudo et al., 2004). This stress activates the HPA axis, with increased titers of the stress hormones ACTH (adrenocorticotropic hormone) and corticosterone (figure 5.6C). Restrained germ-free mice display significantly elevated titers of these stress hormones relative to conventional mice, and this heightened stress response was reversed by colonizing the GF mice with the gut bacterium *Bifidobacterium infantis* (Sudo et al., 2004), suggesting that certain gut microorganisms reduce anxiety under stressful conditions. It is not known precisely how the microbiota influences the HPA axis, but indications that the vagus nerve (the principal route of gut-brain communication via the nervous system) may be important comes from evidence that direct stimulation of this nerve elicits corticosterone release.

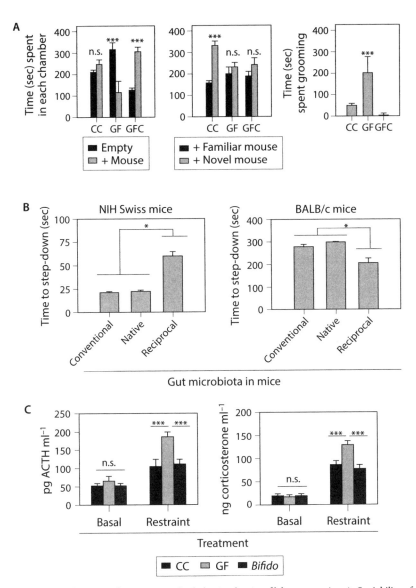

FIG. 5.6. Effect of gut microbiota on complex behavioral traits of laboratory mice. A. Sociability of conventional (CC), germ-free mice (GF), and germ-free mice recolonized with the conventional microbiota (GFC) in three-chamber assays that offer the choice between an empty chamber and chamber with a mouse (left) and between a familiar mouse and novel mouse (middle), and as quantified by time spent grooming (right). B. Step-down latency in GF-mice of two strains colonized with the native gut microbiota (native) and microbiota of the other strain (reciprocal), with SPF mice as untreated control. Mean ± s.e. (15–43 reps). C. Stress hormones titers in conventional mice (CC), germ-free mice (GF), and GF-mice colonized with *Bifidobacterium infantis* (Bifido). Mean ± s.e. (18–24 reps). (Redrawn from Fig. 1b, 1e, and 1h of Desbonnet et al. [2014], Fig. 5C of Bercik et al. [2011], and Fig. 3 of Sudo et al. [2004].)

This brings us to the important question of the mechanistic basis of microbiome-gut-brain axis: what are the microbial products that influence the gut-brain axis, and the host cell types (in gut or brain) with which they interact? This topic is addressed in the next section.

5.3.3. MICROBIAL EFFECTS ON NEURAL PATHWAYS

Studies on the laboratory mouse indicate that neural connections between the gut and CNS are required for some effects of gut microorganisms on the brain. The key nerves innervating the gut are the vagus and spinal nerve; within these nerves, the neurons that communicate information from the gut to the CNS are known as sensory afferents. The mammalian gut also has an enteric nervous system, which comprises neurons that are entirely restricted to the gut wall and includes a large number of sensory intrinsic primary afferent neurons (IPANs), e.g., an estimated 100 million in the human gut. Many IPANs have synaptic connections to the vagal afferents, indicating that the enteric nervous system is exquisitely poised to detect changes in the chemistry of the gut wall, including those induced by changes in microbial activities, and then to communicate these changes to the CNS. In this way, the enteric nervous system is both "the brain in the gut" and a key player in the gut-brain axis.

Neural pathways are an important route for communicating changes in the status of the gut microbiome to the brain. The principal evidence comes from a two-part experimental design, usually conducted on the laboratory mouse. The first treatment comprises mice fed on a suspension of bacterial cells, resulting in changes to brain chemistry that are absent from controls fed on saline. In the second treatment, a subdiaphragmatic vagotomy is performed, interrupting the pathway from the intestine to vagal sensory nuclei, prior to administration of bacteria. If the effect of the bacteria on brain chemistry is abolished for the vagotomized mice, the microbial effect can be attributed to signal transmission along the afferent vagus. This approach has been adopted in studies on the probiotic bacterium *Lactobacillus rhamnosus* strain JB-1™ (Bravo et al., 2011). When this bacterium was administered to mice, expression of the gene coding a receptor of the neurotransmitter GABA (specifically GABA$_{B1b}$) was significantly altered in multiple regions of the brain, including the amygdala, hippocampus, and cortex. Consistent with abundant evidence that GABAergic neurotransmission plays a crucial role in mammalian behavior, including the response to stressful situations, the effect of *L. rhamnosus* on the GABA$_{B1b}$ gene expression was associated with reduced levels of the stress hormone, corticosterone, in the blood and reduced anxiety-like behavior in behavioral tests. However, none of these correlated effects of *L. rhamnosus* on brain function, hormonal status, and behavior were displayed in vagotomized mice, indicating that the introduction of the bacterium to the gut was communicated to the brain via the vagal nerve. Remarkably, the effect of *L. rhamnosus* on the activity of vagal afferents was evident within 10–15 minutes of administering the bacterial cells in ex vivo small intestine preparations (Perez-Burgos et al., 2013), suggesting that the host response is sensory rather than mediated through *L. rhamnosus* effects on the composition of the microbiota or available food sources. Furthermore, these effects are specific to certain bacterial strains: some bacterial strains have neuroactive and psychoactive properties, but many do not.

5.3.4. MICROBIAL PRODUCTS UNDERPINNING THE MICROBIOME-GUT-BRAIN AXIS

Gut microorganisms release a variety of bioactive compounds that may interact directly with cells in the gut, for example, modulating the production of hormones or neurotransmitters by enteroendocrine cells and cytokine production by immune cells, as well as the activity of neurons in the enteric nervous system. Many microbial metabolites are routinely recovered, often at appreciable concentrations, in the plasma and urine of humans and other mammals (Nicholson et al., 2012), indicating that they are transferred into the circulatory system and, thence, throughout the animal body, including to the brain. However, this dissemination of metabolites is constrained by two epithelial barriers, the gut epithelium and the blood-brain barrier, where close apposition of adjacent cells via tight junctions prevents the uncontrolled passage of metabolites between the cells (figure 5.7).

Several classes of microbial metabolites have been implicated as modulators of the gut-brain axis. SCFAs released from microorganisms in the mammalian gut are taken up by gut epithelial cells and, despite their use as a respiratory fuel in the epithelial cells, a proportion of these metabolites escape into the bloodstream and can cross the blood-brain barrier. For example, when mice were fed on ^{13}C-labeled dietary fiber, the gut microbiota fermented the fiber to SCFAs, including acetic acid. ^{13}C-acetic acid was translocated from the gut lumen via the blood to the brain, where it crossed the blood-brain barrier and accumulated preferentially in the hypothalamus (Frost et al., 2014). Within the hypothalamus, the acetic acid is actively metabolized to glutamate, glutamine, GABA, and lactate in astrocytes and neurons, altering neurotransmission by glutamatergic and GABAergic neurons.

A second class of gut-derived compounds with pervasive effects on brain function and behavior are tryptophan metabolites, especially

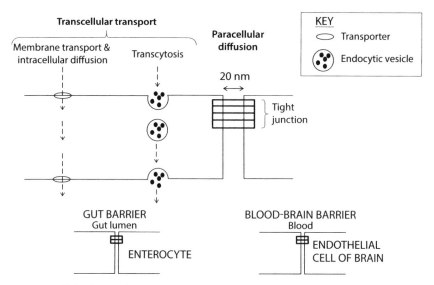

FIG. 5.7. Cellular barriers limiting the spread of microbial products across the gut-brain axis. Transport between cells (paracellular transport) is limited by tight junctions that girdle the lateral margin of every barrier cell, and transport across cells (transcellular transport) is restricted to passive diffusion (e.g., of oxygen, carbon dioxide, and some small lipid-soluble molecules) and selective transport of key metabolites via membrane-borne transporters and selective endocytosis. The gut barrier (bottom left) comprises enterocytes that make up the gut epithelium, and the blood-brain barrier (bottom right) comprises endothelial cells that separate blood capillaries from the extracellular fluid of the brain. This figure displays the relationship for mammals; other animals have broadly equivalent cellular barriers bounding the gut and CNS.

kynurenines. Members of the gut microbiota can synthesize tryptophan and various bioactive derivatives of tryptophan, but their effects on brain chemistry appear to be mediated largely through their effects on tryptophan metabolism by cells of the gut, and not the biosynthetic capabilities of the microbiota per se. Let us consider kynurenine. This metabolite and its derivative 5-hydroxy-kynurenine are released from the gut to the blood, and pass freely across the blood-brain barrier, where brain cells metabolize them to neuroactive metabolites, especially kynurenic acid in astrocytes (shown in figure 5.8) and quinolinic acid in microglia. These compounds bind to receptors of neurons, especially glutamate receptors and the α7-nicotinic acetylcholine receptor, impairing cognition when at high concentrations. An important source of circulating kynurenine is immune cells, especially macrophages, which patrol the gut wall. Kynurenine synthesis by these immune cells is influenced by many factors, including growth factors, sterol hormones, and the profile of cytokines. In particular, kynurenine production is elevated in response to proinflammatory cytokines, whose production is critically dependent on the composition of the gut microbiota: some microbial communities that are proinflammatory tend to increase blood kynurenine titers, while communities dominated by taxa that dampen the gut immune responses tend to decrease blood kynurenine levels (figure 5.8). The cytokine-induced effects of kynurenine and related metabolites on brain function operate in conjunction with direct effects of these cytokines on neurotransmission. Moderate levels of these proinflammatory cytokines and the metabolites that they promote are required for normal neurotransmission, but high levels underpin the behavioral traits of sickness (including fatigue and loss of appetite), and have also been implicated in depression (McCusker and Kelley, 2013) as considered in section 5.3.5.

The gut microbiota is additionally a major source of cell debris, including cell wall fragments, microbial nucleic acids, etc. These products are collectively known as microbe-associated molecular patterns (MAMPs), and their interactions with neurons and immune cells of the gut play an important role in shaping the immunological balance of the gut (see Forsythe et al., 2016;

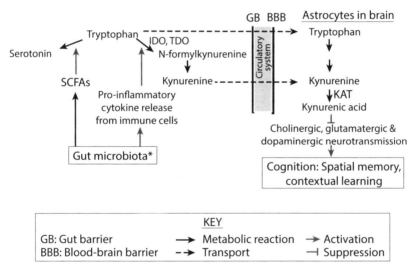

FIG. 5.8. Gut microbiota impacts on brain function mediated by metabolites delivered from the gut to the brain via the circulatory system. Certain gut microorganisms stimulate immune cells associated with the gut to produce proinflammatory cytokines, which activate tryptophan metabolism to kynurenine via the enzymes IDO (indoleamine 2,3-dioxygenase) and TDO (tryptophan 2,3-dioxygenase) in dendritic cells and macrophages. Tryptophan and kynurenine escape to the circulatory system and readily cross the blood-brain barrier, are taken up by astrocytes in the brain, and are metabolized to kynurenic acid. Kynurenic acid is an inhibitor of all classes of glutamate receptors and the α7-nicotinic acetylcholine receptor on neurons; and interference with neurotransmission by high kynurenic acid concentrations negatively affects multiple aspects of cognition. *The composition of the gut microbiota plays a crucial role in determining the balance of pro- and anti-inflammatory cytokines produced by gut-associated immune cells; see Chapter 4, section 4.3.1 for details. (For further information, see Schwarcz et al. [2012].)

Mao et al., 2013; and Table 2.1 and chapter 4, section 4.3). Microbial cell wall fragments can also be translocated from the gut to the systemic circulatory system, e.g., the peptidoglycan of bacterial cell walls is routinely recovered from blood plasma (Clarke et al., 2010). Understanding of the effects of these products is dominated by studies of pathogens, but there is increasing evidence that these microbial products also play an important role in regulating complex behavioral traits of healthy animals. Pioneering research in this area relates to microbial drivers of the healthy sleep-wake cycle of humans and rodents, on which cognitive and emotional health depends. Specifically, peptidoglycan monomers recovered from the urine of sleep-deprived human volunteers have a strong sleep-promoting effect when infused into the brain of rabbits (Krueger et al., 1982). The likely mode of action of bacterial peptidoglycan is via stimulation of proinflammatory cytokines, especially IL-1 and TNFα, because inhibitors of these cytokines abrogate sleep induced by the peptidoglycan fragments (Takahashi et al., 1996). These cytokines have been causally associated with the regulation of sleep-wake behavior in rodents, and display diurnal variation in titers in blood samples, peaking at sleep onset, in humans (Imeri and Opp, 2009). Although the details of these interactions are still poorly defined, they do suggest that microbial products, especially derived from the gut microbiota, can play an important role in shaping host regulatory controls over sleep patterns. Interactions between the microbiome and host determinants of circadian rhythm in animals are considered further in chapter 6, section 6.4.4.

5.3.5. THE GUT MICROBIOTA AND HUMAN MENTAL HEALTH

The evidence from rodent studies that the gut microbiota can influence complex behavioral traits has led to the prediction that the gut microbiota may affect human mental health, including mood, social interactions, and overall psychological well-being. It is very difficult to test this hypothesis on humans. One informative approach has been to quantify the effects of probiotic bacteria on noninvasive indices of brain function and self-reported responses to questionnaires. For example, in a brain-imaging study on healthy volunteers, twice-daily consumption of a fermented milk over four weeks appreciably altered the functional magnetic resonance signal in multiple regions of the brain controlling central processing of emotion and sensation (Tillisch et al., 2013). A double-blind study of 55 volunteers administered defined strains of *Lactobacillus helveticus* and *Bifidobacterium longum* or a placebo over 30 days (Messaoudi et al., 2011) revealed positive effects on multiple indices of emotional health, obtained by self-reported questionnaires, and these results were supported by significant reductions of the stress hormone cortisol in the urine (figure 5.9).

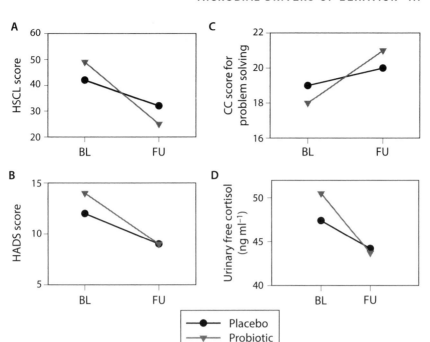

FIG. 5.9. Positive effect of probiotic bacteria on emotional health indices of healthy volunteers. Significant effects of the probiotic (but not placebo) were obtained between the base-line (BL) and follow-up (FU, 30 days later). A. Global severity index for Hopkins Symptom Checklist-90 (HSCL), comprising a questionnaire on psychological distress. The significant reduction in the global index for volunteers administered the probiotic was underlain by significant reductions in anger-hostility, depression and somatization. B. Global score for Hospital Anxiety and Depression Scale (HADS), an alternative measure from HSCL quantifying psychological distress. C. Problem solving capability assessed in the Coping Checklist (CCL), as a measure of cognitive response to adverse events. D. Level of cortisol stress hormone in the urine of the volunteers. Median values (55 volunteers) are displayed. (Figures drawn from data in Messaoudi et al. [2011].)

If probiotic bacteria improve the mental health of healthy volunteers, can they also be used to treat depression and anxiety disorders? These psychiatric conditions tend to be associated with hyperactivity of the HPA axis and proinflammatory cytokine titers, and both of these traits are promoted by perturbations of the gut microbiota in rodent models (see section 5.3.2). There is an expectation, therefore, that microbial therapies would ameliorate these psychiatric disorders. However, a rational approach based on correcting defined microbial perturbations in patients is confounded by the complexity and variability of the human gut microbiome. Although the composition of the microbiota in fecal samples from patients with depression tends to differ from healthy controls, the results vary across different studies, precluding generally applicable conclusions (Forsythe et al., 2016). The data do not support the popular notion of well-defined "happy" and "sad" gut microbial communities.

Large, randomized trials are needed to test whether microbial therapies have any substantive value as an adjunct or alternative to established antidepressant treatments, and understanding of the neuronal mechanisms that might underlie microbial modulation of mood is essential to develop rational microbial therapies.

5.4. Microbes and Animal Communication

5.4.1. THE FERMENTATION HYPOTHESIS

Various studies in recent years have invoked microorganisms as the source of chemical signals mediating communication between conspecific animals. Many of these empirical studies conform to the expectations of the fermentation hypothesis, which proposes that odors used in animal communication are metabolites generated by microbial fermentation of complex macromolecules of animal origin (Ezenwa and Williams, 2014). The fermentation hypothesis was formulated in the 1970s in relation to odoriferous secretions from the anal glands of carnivorous mammals, and this is the focus of this section.

The basis of the fermentation hypothesis is that microbial fermentation products are very diverse and have strong and distinctive odors. We have already considered one important class of microbial fermentation products in this chapter: SCFAs including acetic acid, which are carboxylic acids with up to five carbon atoms (C_1-C_5) produced by microbial fermentation of polysaccharides. The SCFAs are just one group of the total diversity of C_1-C_5 volatile carboxylic acids (VCAs) produced by microbial fermentation of complex polysaccharides, lipids, and proteins. Other VCAs may be unsaturated, bear hydroxyl group(s), have more than one carboxylic acid, etc. Overall, the total profile of fermentation products produced by a microbial community is usually complex and determined by the composition of the substrate, the metabolic capabilities of the different microbial cells, and the patterns of metabolic cross-feeding between the different microbial taxa. As human practitioners in fermentation-based industries (e.g., wineries, production of chocolate, and fermented cheeses and vegetables) can testify, fermentation by complex microbial communities can yield an essentially limitless range of compounds, and animal chemosensory systems (odor and taste) are exquisitely attuned to detect these compounds.

A key study underpinning the fermentation hypothesis was the demonstration that the odoriferous secretions from the anal glands of the Indian mongoose *Herpestes auropunctatus* are dominated by VCAs, the composition of which varies among individuals, and that an individual mongoose can discriminate between different blends of the VCAs (Gorman, 1976). The secretions from the anal glands play an important role in the social interactions of the mongoose, particularly in marking out an individual's home range.

Gorman hypothesized that the VCAs are produced by the dense microbial communities that inhabit the glands and ferment the sebum and other complex compounds secreted into the glands. More specifically, he hypothesized that the unique microbial metabolism in each animal confers a personal odor blend, enabling individual recognition. In this way, the fermentation hypothesis of individualized communication among carnivorous mammals prefigured the later proposition that the gut microbiota is "as unique as our fingerprints" (chapter 3, section 3.2.2).

We can take the reasoning a step further than considered by Gorman. Not only is the profile of microbial fermentation products unique to each individual animal, it is also potentially informative of the condition and status of the animal. The developmental age, sex, reproductive status, and health status of the animal are likely to influence both the composition of secretions into the anal glands that the microbiota ferments, and also (via immune and other factors) the composition of the microbiota. If this is correct, then the composition of the released fermentation products is not only the basis for individual and group recognition, but also provides generic information about the status of the animal. Speculating further, the profile of fermentation products may be a reliable indicator of host status. For example, the wide array of substrates for microbial fermentation and the immune-surveillance required to maintain a diverse microbiota may be costly, such that an animal in poor health cannot support microbial production of a complex array of fermentation products that is indicative of good health. In other words, the microbial fermentation products may be honest signals of the status and quality of the animal host.

Anal glands occur in all mammals of the order Carnivora, including the mongooses and meerkats, the cats, hyenas, and civets, and the bears, dogs, and raccoons. Their odoriferous volatile secretions play important roles in the social interactions of these mammals, including individual and group identity, territory marking, and recognition of mate and offspring. Where studied, microorganisms are detected in the anal glands, and VCAs in the secretions (Ezenwa and Williams, 2014). The most comprehensive study to date concerns hyenas, whose anal glands are richly colonized by a diverse community of fermentative microorganisms, including Firmicutes, Actinobacteria, and Bacteroidetes (Theis et al., 2012). As predicted for the fermentation hypothesis, both the composition of the microbiota and the profile of volatile fermentation products vary among individual animals, for both of the two hyena species studied, the striped hyena *Hyaena hyaena* and spotted hyena *Crocuta crocuta* (Theis et al., 2013). Furthermore, microbial composition and their fermentation products covary with sex and reproductive status of the host, consistent with the hypothesis (above) that the microbial fermentation products may be honest signals of host status and quality.

The various studies on the secretions from the anal glands of carnivorous mammals are highly suggestive of microbe-mediated communication. However, definitive evidence to demonstrate that volatiles mediating animal communication are synthesized by microorganisms requires multiple lines of evidence (Douglas and Dobson, 2013). These should include the demonstration that the microorganism(s) produce the compound; that the volatile and associated effect on the behavior of other animals are abrogated in animals lacking the microorganisms; and that the chemical and behavioral responses are recovered by reintroducing the microorganisms back to the animal. This quality of experimental evidence would be exceptionally demanding (and arguably impossible) to achieve with free-ranging mammalian carnivores, and the evidence that these animals have outsourced the production of "info-chemicals" to microorganisms remains as highly suggestive, rather than conclusive.

In the following two sections, the fermentation hypothesis is pursued in two different ways. First (section 5.4.2), we address one requirement of the hypothesis, that the microbial volatiles within one animal do not vary stochastically (which would preclude reliable identification of the host by the microbial volatiles) and are more similar to other members of the same group than to members of different groups. Then (section 5.4.3), we consider independent research on the role of microbial fermentation products in mediating insect behavior, with conclusions that are remarkably like the predictions of the fermentation hypothesis in mammals.

5.4.2. SOCIAL INTERACTIONS AND THE DISTRIBUTION OF MICROORGANISMS

Microbial fermentation products have been proposed to be used by various mammalian carnivores as signals of group identity. Consistent with this hypothesis, the composition of the bacterial communities associated with the anal glands of the spotted hyena varies to a significantly greater extent between different groups than within groups among free-ranging populations in Kenya (Theis et al., 2012). Within-group similarity may be promoted by microbial transfer, mediated by the frequent contact between individuals of the same group and also by the distinctive overmarking behavior of this species, which involves members of the same group depositing a paste of anal gland secretion at the same spot, often in quick succession. These suggestions are very much in accord with data on other mammals, most particularly primates, indicating that social group is an important predictor of the composition of the microbiota.

Among primates, the gut microbiota in chimpanzees in Gombe Stream National Park in Tanzania is patterned more strongly by social group than by genetic relatedness, age, or sex (Degnan et al., 2012). Similarly, the variation in the gut microbiota within members of one human family is reduced relative to nonfamily members in the wider community (Schloss et al., 2014), and shared residence is also a strong predictor of the skin microbiota in humans (Song et al., 2013). In principle, many factors, including genetic relatedness, shared diet, and transmission by contact, may contribute to these patterns, but there is increasing evidence that social interactions play a primary role. For example, analysis of the gut microbiota in adult baboons in two social groups of the Amboseli Baboon Research Project (Tung et al., 2015) confirmed that, as in other primate (including human) studies, social group accounted for an appreciable proportion of the among-individual variation in micro-biota composition. Furthermore, detailed behavioral observations revealed that grooming partner frequency was the chief determinant of between-individual similarity of gut microbiota. It is very likely that microorganisms were transferred between grooming partner, and then gained access to the gut by frequent hand-to-mouth contact.

The importance of these studies in relation to the use of microbial prod-ucts for communication of group identity is that they provide a reinforcing mechanism for maintenance of within-group similarity of microbial function: specifically, a positive feedback loop of increasing social interaction and in-creasing similarity of microbial community composition and function (figure 5.4B). These considerations raise the possibility that the role of microbial products in individual recognition and group identity may not be restricted to mammalian carnivores, but are more widely distributed. To take one spe-cific example, individual recognition in the house mouse is likely mediated by a large family of urinary proteins (major urinary proteins, MUPs), which bind to a great diversity of volatiles (Hurst et al., 2001). As an extension of the fermentation hypothesis, it is possible that some of these volatiles may be of microbial origin, for example synthesized by the gut microbiota, and then circulated in the blood stream and delivered to the urine.

5.4.3. MICROBIAL FERMENTATION PRODUCTS AS DRIVERS OF INSECT SOCIAL BEHAVIOR

Section 5.4.1 introduced the hypothesis that mammalian carnivores may use microbial fermentation products in communication first, because these products are diverse with small differences in composition readily detected by the animal chemosensory systems, making them well-suited for indi-vidual or group recognition and second, because the composition of the fermentation products may be honest signals of the health and reproductive status of the animal host. In section 5.4.2, it was argued, further, that social interactions among members of a group promote within-group transmission of microorganisms, thus promoting the reliability of microbial-mediating signaling of group identity. Similar conclusions are emerging from studies on insects, with the additional advantage that these insect systems are more amenable than free-ranging mammals to experimental manipulation.

One persuasive study concerns the German cockroach, *Blatella germanica* (Wada-Katsumata et al., 2015). These insects aggregate, attracted by VCAs released from their feces. In olfactometer experiments, larval cockroaches offered a choice between feces from conventional and axenic *B. germanica* display a strong preference for the conventional feces (figure 5.10–1), implicating the gut microbiota in the behavioral response. Parallel chemical analysis revealed that 31 of the 40 VCAs detected in the feces from conventional cockroaches were undetectable or much-reduced in the axenic feces; and the cockroaches were significantly attracted to a blend of 6 abundant VCAs in the conventional treatment (figure 5.10–2). They were also attracted to a paste of bacteria isolated from the cockroach gut relative to axenic feces (figure 5.10–3), further implicating the bacteria as the source of behaviorally important VCAs. However, the greater responsiveness to conventional feces than the bacterial paste suggests that many bacteria, possibly including

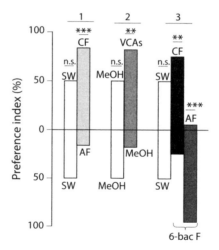

FIG. 5.10. Orientation of the German cockroach *Blatella germanica* towards fecal extracts in a two-choice olfactometer assay. 1. Preference for conventional fecal extract (CF) over axenic fecal extract (AF) in sterile water (SW). 2. Preference for a synthetic blend of volatile carboxylic acids (VCAs: isovaleric acid, valeric acid, succinic acid, benzoic acid, phenylacetic acid, 3-phenylpropionic acid) dissolved in methanol over methanol (MeOH). 3. Preference for fecal extract from cockroaches colonized with 6 bacteria (6-bacF, comprising *Enterococcus avium*, *Weissella cibaria*, *Pseudomonas japonica*, *P. monteilii*, *Acinetobacter pittii*, and *Acinetobacter* sp.), relative to CF and AF. Statistical analysis was n.s. (not significant), * p<0.05, ** p<0.001, or ***p<0.001. (Redrawn from Fig. 2C, Fig. 4, and Fig. 5 of Wada-Katsumata et al. [2015].)

unculturable taxa, or microbial metabolism of specific substrates in the feces may be required to replicate the full insect response.

The fitness of the German cockroach in the natural habitat is critically dependent on aggregation, which confers protection against desiccation and predators, as well as promoting mate location in adults. Why outsource this vital function to microorganisms? Likely explanations are similar to the arguments relating to fermentation products of the anal gland microbiota of mammalian carnivores (section 5.4.1). The gut microbiota of the cockroach is predicted to vary with habitat and insect diet, and to be very similar among group members of different genotype through among-insect transfer. Consequently, the specific blend of volatiles produced by the microorganisms may be a more reliable aggregation signal than a signal of insect origin that reflects insect genotype. If cockroaches learn the volatile blend of their group, the spatiotemporal variation in the VCAs produced by the microbiota could potentially promote exquisite specificity of the aggregation response. Also, similar to the argument that microbial volatiles may be honest signals in mammals (section 5.4.1), the predicted variation in the volatile blend with diet and composition of the microbiota may provide information on the quality of the food (e.g., is it nitrogen-rich?) and microbiota (e.g., is it free of pathogens?) in a cockroach group. In this way, a pheromone of microbial origin may provide accurate information supporting critical behavioral decisions about whether to aggregate or disperse, and whom to aggregate with. We return to consider some evolutionary consequences of outsourcing function to microorganisms in chapter 7 (section 7.2.3).

5.4.4. BEYOND THE FERMENTATION HYPOTHESIS: SIGNAL EVOLUTION BY HOST CAPTURE OF MICROBIAL METABOLISM

In addition to the developing evidence for a role of microbial fermentation products in animal communication, other data indicate that nonfermentative products of microbial metabolism may also influence social interactions among animals. One example is trimethylamine (TMA) in the house mouse, produced by gut microbiota from dietary choline and excreted into the urine (Li et al., 2013). TMA production in mice is of particular interest because it is strongly sex-specific, with considerably higher concentrations in the urine of males than females (figure 5.11A). This difference does not result from any difference in the microbiota between the sexes, but can be explained by the much-reduced expression of a key gene, flavin-containing monooxygenase 3 (*FMO3*), which oxidizes TMA in the liver of male mice, relative to females (figure 5.11B). Furthermore, this sex difference is not general among mammals (as discussed in chapter 3 (section 3.4.3), TMA is oxidized efficiently in both

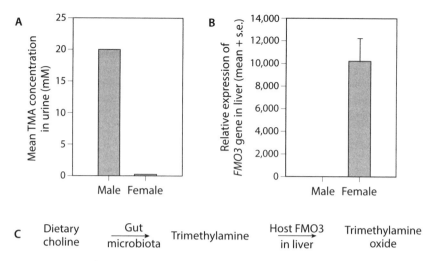

FIG. 5.11. Sex-specific production of trimethylamine (TMA) in the mouse. A. The concentration of TMA in the urine of adult mice. B. Expression of the flavin-containing monooxygenase 3 gene (*FMO3*) in the liver of 6-week-old mice. C. The TMA titer in the mouse is determined by the balance of synthesis by the gut microbiota and degradation by FMO3 in the liver. (Redrawn from text, Fig. 1D, and Fig. 2F of Li et al. [2013].)

men and women). Instead, the suppression of *FMO3* expression in adult males evolved in one lineage within the genus *Mus*, including the house mouse (Li et al., 2013). These findings reveal an important general point, that although the impact of microbial products on animal behavior is often dictated by the metabolic function of the microorganisms, some microbial-mediated signaling systems may evolve through changes in expression patterns of key host genes.

5.5. Summary

Animals behave: they perform evolutionarily adaptive actions in response to environmental stimuli, as influenced by their internal state. Contrary to the traditional view that the behavior of a healthy animal can be explained entirely by the function of the nervous system in conjunction with the endocrine and immune systems, there is now increasing evidence that animal-associated microorganisms influence multiple behavioral traits, especially in relation to feeding behavior, mental well-being, and animal communication.

The chief evidence for microbial influence on animal feeding comes from the effect of microbial perturbations that alter food consumption (section 5.2). In particular, mice with a null mutation in the *TLR5* gene display hyperphagia, and microbial involvement is indicated by hyperphagia in wild-type mice colonized with the gut microbiota from the mutant mice. It has been suggested that microorganisms may alter the function of animal signaling circuits that regulate feeding by synthesizing signaling molecules, thereby increasing the titer of appetitive or satiety signals, or by producing metabolites that alter the production of these signaling molecules in the animal. Furthermore, these microbial effects may be more pronounced in hosts with microbial communities of low diversity, where the dominant taxa are very abundant, although this requires experimental validation.

In a similar fashion, some of the most persuasive evidence for microbial effects on mental well-being comes from microbiome perturbation studies (section 5.3). Germ-free mice display behavioral traits indicative of altered anxiety or social interactions, as well as learning capabilities, and these differences have been associated with changes in neuronal function or titers of neuroactive metabolites in key regions of the brain. Complementary neurophysiological and endocrine data indicate that members of the gut microbiota modulate communication between the gut and the brain, either via the vagus nerve or by impacts on bioactive compounds circulating in the blood. The relevance of these studies on rodent models to humans is suggested by the results of double-blind trials on healthy volunteers in which probiotic bacteria improved self-reported indices of emotional health.

A valuable framework for investigating the role of microorganisms in animal communication is provided by the fermentation hypothesis, originally developed to explain the olfactory basis of social behavior in mammalian carnivores (section 5.4). It is argued that the odoriferous volatiles that are released from the anal glands and function in the recognition of individual animals and group members are the products of microbial fermentation of complex lipids and other substrates in the glands. The fermentation hypothesis is supported by evidence that the composition of the microbial communities in the anal glands of hyenas vary more between animals from different groups than from the same group, and this taxonomic variation is correlated with variation in the volatile fermentation products released from the glands. Similar studies conducted on an insect, the German cockroach, demonstrate that the aggregation pheromone of this species is a complex mix of fermentation products synthesized by the gut microbiota. Microbial fermentation products offer exquisite specificity (suitable for individual or group recognition) because different microbial communities generate different blends of compounds, which can be discriminated very readily by the chemosensory systems of animals. The composition of the fermentation products may also be reliable indicators of the health and reproductive status of the animal, although this remains to be established.

More generally, many metabolites in the body fluids of animals are the products of animal-associated microorganisms, offering the opportunity for the evolution of novel signaling molecules by coupled microbial-host metabolism. For example, the male-specific pheromone, trimethylamine, mediating social communication in the house mouse arises from differences between the sexes in metabolism of this microbial product.

Study of the microbial drivers of behavior in the healthy animal is a very young discipline, currently generating many more review articles than original research papers. To date, the empirical data often provide incomplete and occasionally contradictory evidence, suggesting that the microbial effects are complex and often subtle. We should not anticipate a simple microbial explanation for complex behavior or a microbial cure-all for the many behavioral and psychiatric disorders that beset humankind. Nevertheless, there is now a sufficient body of evidence to indicate that animal behavior is influenced not only by pathogens and parasites, but also by the microbial communities that inhabit all healthy animals.

6

The Inner Ecosystem of Animals

6.1. Introduction

Until recently, ecology was not at the forefront of discovery in microbiome research. The transformation of our understanding of animals from a unitary entity of genetically uniform animal cells to a multiorganismal animal-microbial system has been driven largely by the disciplines of microbiology, immunology, and biomedical science. However, the situation is changing very rapidly as microbiome research moves from describing the taxonomic composition and gene content of microbiomes to understand how microorganisms interact with each other and their animal host. It is increasingly recognized that an ecological framework offers a valuable route to explain and predict these complex interactions (Costello et al., 2012; Coyte et al., 2015). In essence, an animal and its microbiota comprise an ecosystem: an inner ecosystem that is nested within the outer ecosystem, in which the animal host interacts with external organisms and the physical environment (figure 6.1A).

Many of the principles of ecology command broad consensus, but some very basic problems remain. One of the most fundamental dichotomies in ecology has been a debate that was brought into sharp focus in plant community ecology during the 1920s. Are plant communities so tightly structured and predictable that they can be considered as single entities or "super-organisms," as argued by Clements; or are communities the product of a myriad of individual interactions between organisms and their environment, in combination with chance historical events, as propounded by Gleason? Plant ecologists in the twentieth century largely resolved the Clements vs. Gleason dichotomy, broadly favoring Gleason, but some statements in the recent microbiome literature are echoes of this historical debate. For example, the

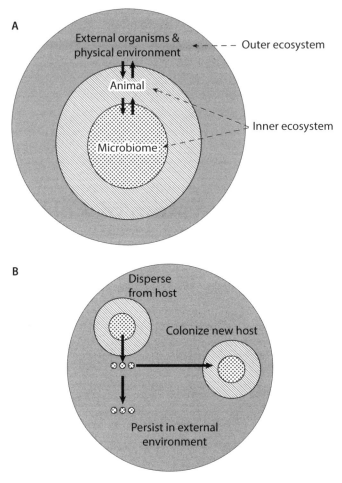

FIG. 6.1. Animals as ecosystems. A. The inner ecosystem of an animal and its microbiome, nested within the outer ecosystem of the external environment. Many microorganisms and microbial communities are entirely enclosed within the animal as shown. Some animals bear a surface microbiota (known variously as ectosymbiotic and epibiotic communities, e.g., skin microbiota of humans), reversing the host-microbial relationship, such that the microbial partners mediate the interaction between animal and outer ecosystem (not shown). B. Animal-associated microorganisms disperse between animals, as well as between animals and the external environment, generating a global population (known as a metapopulation) that is transmitted across multiple subecosystems (also known as patches).

assertions that the microbiome is an organ and a human is a superorganism without empirical evidence are simplistic Clementsian claims. Other ecological concepts, including the meaning of ecological stability and how to measure it, the processes underlying the relationship between taxonomic diversity and ecosystem stability, and the role of contingency in shaping ecological communities, are the subject of intense debate. Ecological interpretations of microbiome data need to be aware of the wider uncertainties about the underlying processes and consequences of diversity, community structure, and ecosystem function.

This chapter addresses three topics of increasing ecological complexity, from the ecology of single taxa through ecological communities to ecosystem processes. Section 6.2 considers the abundance and distribution of individual microbial taxa, focusing on the patterns of their distribution between animal hosts and the free-living environment, together with the underlying processes. Section 6.3 focuses on ecological communities, particularly the contribution of among-microbe interactions and host-microbe interactions in shaping the taxonomic and functional composition of microbial communities associated with animals. The perspective is broadened further in section 6.4 to address the properties of the totality of the microbial communities and animal host as an ecosystem. The properties of key importance are the determinants of ecosystem stability, on which animal health and fitness depend, and processes that shape successional changes in the microbiota as the animal host develops and ages, and in response to major perturbations.

6.2. The Abundance and Distribution of Animal-Associated Microorganisms

6.2.1. ANIMAL-ASSOCIATED MICROORGANISMS WITH LIMITED OR NO FREE-LIVING POPULATIONS

Let us consider the nested ecosystem concept illustrated in figure 6.1A further. Many of the organisms in the external environment with which our focal animal and its inner ecosystem interact also bear microbial communities. This means that the outer ecosystem contains multiple subecosystems (figure 6.1B), and, because animals die, each one of these subecosystems is ephemeral. The ecological scope of animal-associated microorganisms depends on their capacity to (1) disperse from the current host, (2) colonize other hosts, and (3) persist as free-living populations in the external environment independent of animals. Because microorganisms disperse between spatially separate patches (which may be animals or the free-living environment), the perspective of a single animal is inadequate to address the global abundance and distribution of animal-associated microorganisms. Instead, it is useful to treat the microbial partner as a metapopulation, comprising the total group of these interacting populations (figure 6.1B).

The significance of the animal as determinant of the distribution of microorganisms is most evident for microorganisms localized exclusively to their animal hosts (figure 6.2A). With no capacity to disperse into the external environment, these microorganisms are, metaphorically, trapped in a gilded cage. This type of relationship is displayed in various invertebrate animals with bacteria that are housed in specialized animal cells (bacteriocytes) and are vertically transmitted to the host offspring. In insects, where these associations have been studied in most detail, the bacterial cells are transferred from the bacteriocytes to developing oocytes in the female ovary, persist through fertilization and embryogenesis, until the bacteriocytes differentiate in the embryo and incorporate the bacteria. As discussed in chapter 7 (section 7.3.2), exclusive vertical transmission over many host generations leads to massive genome reduction of the bacterial partner, such that they are unable to survive apart from the insect host. Other microorganisms that are restricted to the animal habitat include obligate anaerobic microorganisms in the anoxic regions of animal guts. Because the great majority of animals live in oxic environments, the persistence of oxygen-intolerant gut microorganisms requires rapid transfer between animal hosts. For example, termites bear a complex community of obligate anaerobic microorganisms in the anoxic chamber ("paunch") of their hindgut. These microbes are lost when the termite molts, a process that introduces oxygen into the hindgut, and each termite is recolonized postmolt by feeding on microbe-rich droplets released from the anus of other colony members, ensuring that the microbial symbionts have fleeting exposure to the oxygen-rich external environment.

For some animal-associated microbial taxa, viable cells are shed into the external environment, generating free-living populations. However, these free-living populations may very commonly be "sink populations," i.e., individual cells have low survivorship or proliferation in the external environment, and the free-living population is sustained by continued input from animal hosts (figure 6.2B). For example, humans shed microorganisms into the air, but this microbial "cloud" does not sustain persistent free-living populations (Meadow et al., 2015). Instead, these transiently free-living cells, together with microbes transferred by direct contact, contribute to the similarity of microbial communities in people, and even their pets, living in the same household (Song et al., 2013).

In some animals, microbial shedding has evolved into specialized mechanisms for transmission of microorganisms. This is illustrated by insect bugs of the family Plataspidae. The hindgut of these insects bear a unibacterial culture of the γ-proteobacterium *Ishikawaella*, which is unculturable and unknown apart from their insect host. *Ishikawaella* cells are transmitted from mother to offspring via specialized fecal pellets, often known as symbiont capsules that are resistant to desiccation and other abiotic factors (Fukatsu and Hosokawa, 2002). Specifically, the adult female deposits eggs onto the leaf of a plant, and she regularly alternates between oviposition behavior and the deposition of a symbiont capsule from the hindgut; and, immediately on hatching, each offspring insect extends its proboscis into the adjacent symbiont

A. No free-living population: Direct transfer from one host to the next

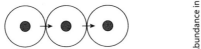

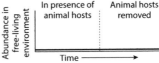

B. Free-living sink population: Sustained by shedding from animal hosts

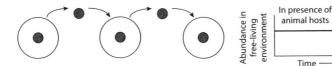

C. Self-sustaining free-living population, independent of input from animal hosts

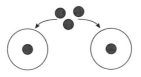

D. Free-living population: Sustained by animal-associated population

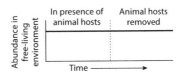

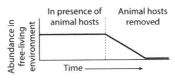

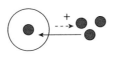

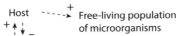

FIG. 6.2. Distribution of microbial populations between animal hosts and the external environment (host: large white circles; microorganisms: small grey circles). A. Free-living populations absent, e.g., in associations with obligate vertical transmission. B. Free-living populations sustained by input from animal hosts; these populations may contribute to the horizontal transmission of microorganisms between hosts. C. Persistent free-living populations, independent of the presence of animals. D. Free-living population is promoted (dashed arrow) by beneficial effects of the animal-associated population on animal abundance or activity (top), creating a positive feedback loop, even under conditions where the animal-associated microorganisms may comprise a sink population (bottom).

capsule and sucks up an inoculum of the bacterial cells, so establishing its gut microbiota. Another intriguing candidate instance relates to the luminescent *Vibrio fischeri* that colonizes the light organ of the squid *Euprymna scolopes*. This bacterium attains appreciable populations in the water column of the

E. scolopes habitat, but is otherwise rare or undetectable in the external environment (Lee and Ruby, 1994). The free-living populations are likely sustained by the daily expulsion of >90% of the *V. fischeri* from the squid light organ (Wier et al., 2010), but the fate of the cells in the water column has not been investigated. As well as being a sink population, the free-living cells may be the source of inoculum for young squid individuals.

6.2.2. ANIMAL-ASSOCIATED MICROORGANISMS WITH SUBSTANTIAL FREE-LIVING POPULATIONS

In contrast to the microorganisms described in section 6.2.1, some micro-organisms in animal microbiomes are predicted to sustain free-living popu-lations that are independent of input from animals (figure 6.2C). For these microorganisms, the occasional capture of microbial cells by animals, e.g., via bulk flow of air or water during respiratory exchange or during feeding, may have little or no effect on free-living populations. Indeed, animal-associated populations may be sink populations for some microbial taxa. Nevertheless, animals can have significant effects on the distribution of these microorgan-isms, even where the size and dynamics of free-living populations in one habitat patch are independent of animals. In particular, mobile animals can mediate microbial dispersal over distances orders of magnitude greater than can be achieved by independent movement of microorganisms, and facilitate access to new habitats that cannot readily be reached by abiotic routes, e.g., against water currents or prevailing winds.

By and large, these different population structures can be discussed only in generalities because little is known about the effect of animals on the dy-namics of free-living populations of many microorganisms. Understanding of these processes will require analysis over multiple spatiotemporal scales because the significance of the animal habitat for microorganisms may vary with ecological circumstance. Rare but large advantages of the animal habitat, e.g., as refuge from extreme abiotic conditions or enemy-free space under con-ditions of exceptional predator pressure, may play a deciding role in shaping the abundance and distribution of free-living populations of microorganisms.

We should additionally consider the possibility that animal-associated microorganisms promote their own free-living populations by processes independent of shedding from the host into the environment. As illustrated in figure 6.2D, this situation can arise via a positive feedback loop, where an animal promotes free-living populations of microorganisms that, in association with the animal, favor animal abundance or activity. This feedback loop can operate even under conditions where the animal-associated microorganisms comprise a sink population. Likely ecological scenarios for this type of inter-action relate to the capacity of animals to alter the suitability of the external environment for free-living populations of microorganisms, for example by releasing nutrient-rich substrates (e.g., mucus) or by physical disturbance (e.g., animal burrowing). To my knowledge, these types of interactions have been given little consideration in animal-microbe interactions. However, an equivalent scenario has been proposed for the relationship between leguminous plants and their nitrogen-fixing rhizobia bacteria that inhabit root nodules. Although rhizobia bacteria can persist for extended periods in the absence of plant hosts, their populations are strongly promoted by the presence of compatible plants. It has been debated vigorously whether this host effect can be attributed to the release of viable bacteria from the plants into the soil at senescence of the nodules (figure 6.2B) or the release from rhizobia-colonized roots of compounds that specifically promote free-living rhizobia populations (figure 6.2D) (Bever and Simms, 2000; Denison and Kiers, 2011).

The variety of interactions between animal-associated and free-living populations of microorganisms raises many questions, including the selection pressures on the animal and microbial partners that both shape, and are shaped by, the population dynamics of the microbial partners. For example, are the host-associated microorganisms in figure 6.2D exploited by the host or are they altruists that die for the benefit of their free-living kin? These evolutionary issues are addressed in chapter 7 (section 7.2.2).

6.3. Ecological Processes Shaping Microbial Communities in Animals

6.3.1. TAXONOMIC AND FUNCTIONAL INDICES OF COMMUNITY COMPOSITION

Studies of microbial communities in animals are usually initiated by an analysis of microbial composition and diversity within individual animals, and how these indices vary among animals (figure 6.3A). Very commonly, the microorganisms are characterized exclusively in terms of their taxonomy but, as this section considers, functional indices of composition and diversity can provide very important complementary information.

For taxonomic indices of diversity, the method of choice is to determine the DNA sequence of a region in the rRNA operon (e.g., 16S rRNA gene of bacteria, and the internal transcribed spacer regions (ITS) of fungi), providing insight into the diversity of animal-associated microorganisms independent of whether the microorganisms can be cultured. However, the phylogenetic trees built from these data refer only to the sequences under study, and not necessarily the rest of the genome, including genes that determine functional traits. At the largest phylogenetic scales for bacteria, the 16S trees can resemble trees based on protein-coding gene content, and this correspondence is

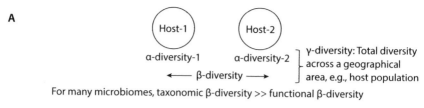

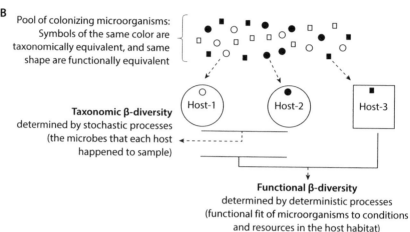

FIG. 6.3. Taxonomic and functional diversity of microorganisms associated with animals. A. α-and β-diversity: α-diversity refers to the diversity within a habitat (e.g., an animal host), and β-diversity refers to the difference in diversity between habitats. Very commonly, the taxonomic β-diversity is considerably greater than the functional β-diversity because many microorganisms are functionally equivalent, i.e., there is functional redundancy in microbial communities. B. Mixed determinants of microbial community assembly. From a functional perspective, assembly is deterministic, dictated by the ecological fit between the functional traits of the microorganisms and the conditions and resources in the host environment. From a taxonomic perspective, the assembly is, at least partly, stochastic, shaped by host sampling from a pool of multiple functionally equivalent microbial taxa.

used widely to infer the gene content (and hence function) of uncultured bacteria in the human gut microbiome from 16S sequences (Langille et al., 2013). However, the correspondence between taxonomy and function is far from perfect. All genomes, especially those of many bacteria, are in continual evolutionary flux, gaining genes by horizontal acquisition from various different organisms, and losing DNA especially under conditions where rapid proliferation and hence small genome size are advantageous (Boon et al., 2014). Consequently, taxa with very similar or identical 16S sequence may be functionally different, while highly divergent taxa can have functionally equivalent traits (well-known examples include the capacity to fix nitrogen or reduce sulfate, which are displayed by some, but not all members, of multiple bacterial phyla). Functional redundancy, i.e., many taxa with equivalent functional traits, is widespread in microbial communities, including those associated with animals.

The opportunity to investigate functional indices of the composition and diversity of microbial communities in animals has come with culture-independent methods to quantify function. For many purposes, the method of choice is metagenomics, i.e., the total complement of genes determined by shotgun sequencing of the total DNA in the sample, which provides an excellent overview of the functional potential of the microbial community. The equivalent analysis of RNA, metatranscriptomics, provides a quantitative index of functional activity at the time of assay. A comparative metagenomics and metatranscriptomic analysis of the human gut microbiome revealed good correspondence between the relative abundance of many genes in the metagenome and metatranscriptome, although the among-individual variation was consistently greater for the metatranscriptome than the metagenome and certain functions tended to be overrepresented (e.g., methanogenesis) or underrepresented (e.g., sporulation) in the metatranscriptomes (Franzosa et al., 2014). To some extent, these methods also provide taxonomic information because the computed sequences of genes or transcripts can be assigned to particular microbial taxa by sequence similarity to genes of taxonomically identified organisms in public databases. Even more informative are the emerging methods of single-cell genomics and transcriptomics, which can be used to infer the functional traits of individual members of the communities.

The capacity to determine the taxonomic diversity (inferred from sequence of the rRNA gene operon) and functional diversity (metagenomics/metatranscriptomics) provides the basis to address the ecological processes shaping the composition of microbial communities in animals (figure 6.3B). It can be argued that, from a functional perspective, the colonizing microorganisms are those that can tolerate the conditions (pH, oxygen tension, temperature, etc.) in the host habitat and can utilize host-derived resources, such as mucus shed into the gut lumen or lipids secreted from the sebaceous glands of the skin. These conditions and resources are the environmental filter that limits the functional diversity of microorganisms colonizing and thriving in the animal. Although other factors (e.g., among-microbe interactions) can also influence community composition, we should expect some predictability in community assembly from a functional perspective. From the taxonomic perspective, however, each individual host is colonized by just a subset of the global pool of functionally-compatible microorganisms, and—because of functional redundancy—different hosts can be colonized, by chance, with different subsets of functionally equivalent microorganisms. In other words, stochastic processes can play a large role in shaping the taxonomic composition of microbial communities in one host and the differences between hosts. This point is illustrated by the tremendous taxonomic β-diversity revealed by early studies on the human gut microbiota: that the microbiota of each person "is as unique as our fingerprints" (Dethlefsen et al., 2007). However, this variation is not necessarily indicative of functional variation, as was subsequently demonstrated by the consistently lower functional β-diversity than taxonomic β-diversity of microbial communities in various sites on the human body (Human Microbiome Project Consortium, 2012a).

The taxonomic and functional indices of microbial communities associated with animals provide valuable descriptions of the communities. However, their greatest value is that they can be used to investigate the ecological processes that shape the structure of microbial communities. The key processes are among-microbe interactions and host-microbe interactions, as are considered next.

6.3.2. AMONG-MICROBE INTERACTIONS

We know that members of microbial communities associated with animals interact. Most studies have focused on competition and mutually beneficial cross-feeding of metabolites, usually in simplified communities containing just two (or a few) taxa (e.g., chapter 2, section 2.2.2). Although many of these studies are very elegant demonstrations of among-microbe interactions, they suffer from the fundamental caveat that they lack ecological context. As Konopka (2009) has written: "although we have textbook examples of these phenomena [among-microbe interactions], the breadth of their ecological significance awaits discovery." The problems are two-fold. The first is the relevance of interactions that have been identified in simplified systems: do they occur and are they important in complex natural communities in animals? The second caveat relates to the trophic structure of the communities. The competitive and mutualistic interactions operate at a single trophic level. Could predation and parasitism play important roles in shaping the microbiome in animals, just as they influence community structure in the external ecosystem of animals and other macroorganisms? The focus on interactions among few taxa at one trophic level has all the advantages of being technically feasible. However, like the man searching for his keys under the lamppost where visibility is good, even though he has no reason to believe that his keys are in this location, studies of the microbiome that focus on a interactions within a single trophic level could be missing critically important interactions.

How can we obtain a global overview of among-microbe interactions? One developing solution to this problem is to analyze the pattern of co-occurrence of different microorganisms in different hosts. Two (or more) microbial taxa are described as co-occurring where they are found together across multiple hosts more often than predicted by chance, while coexcluding organisms have the reverse distribution pattern (figure 6.4A).

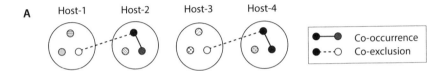

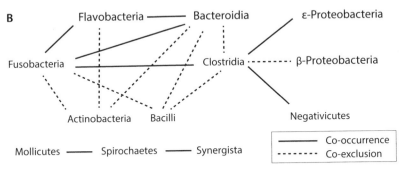

FIG. 6.4. Co-occurrence patterns among microorganisms. A. Identifying co-occurrence (or coexclusion) from correlated (or anticorrelated) patterns of presence/absence across many habitats (hosts). B. Co-occurrence of bacterial taxa (classes) in the microbiome of the human mouth. (Redrawn from Fig. 4 of Faust et al. [2012].)

It is very important to appreciate that co-occurrence patterns may have multiple alternative explanations. Some co-occurring organisms may share similar habitat requirements but not interact (or barely so), while others may interact cooperatively, including displaying mutual dependence, or interact indirectly (e.g., two prey species of a common predator). There is also variation in the functional traits of co-occurring organisms, ranging from high similarity (forming a guild of functionally-equivalent taxa with strong niche overlap) to major differences. Some of the cooperative interactions involve taxa with complementary functions, especially metabolic capabilities, providing the opportunity for metabolic cross-feeding. In the same way, coexclusion may be indicative of major differences in habitat requirements (e.g., between obligate aerobes and obligate anaerobes) or competitive interactions, such as a toxin-producing bacterium that excludes all susceptible taxa, as considered in chapter 4 (section 4.4.2). In many systems, the alternative interpretations of positive and negative interactions can be tested experimentally.

Methods to study co-occurrence patterns are well developed and are beginning to be applied to microbiome research (Berry and Widder, 2014; Faust and Raes, 2012). The distributions of animal-associated microorganisms can be explored as pair-wise correlations of prevalence (presence/absence, as illustrated in figure 6.4A) or abundance, as well as multiway comparisons using regression techniques. These data can then be visualized and investigated as networks of positive and negative correlations. For example, the bacterial communities in the human mouth can be described as multiple positive and negative correlations among different taxonomic groups, displayed in figure 6.4B. This network is driven, at least in part, by the tendency for closely related taxa to co-occur (probably the result of niche overlap) and for distantly related taxa with similar functional traits to coexclude each other (likely through competition) (Faust et al., 2012). This analysis of the mouth microbiome formed part of a larger analysis of correlation networks across 18 body sites in healthy people that also revealed weak connections between many sites. In other words, the human body is interacting not with a single interconnected microbiome but with multiple microbial communities that are largely independent of each other.

Despite uncertainties about the biological meaning of positive and negative correlations in co-occurrence networks, their contributions to the overall structure of networks have been discussed in various contexts. As in free-living microbial communities, animal-associated microbial communities tend to be dominated by negative correlations. A long tradition of ecological theory explains this result in terms of community stability: cooperative interactions can cause instability via coupled population increase of the cooperating partners or correlated losses of mutually dependent taxa, while competitive interactions tend to dampen these potentially disruptive positive feedback loops (Coyte et al., 2015). Balancing this interpretation is system-level evidence for positive interactions within subcommunities in diverse microbial assemblages, including the human gut (Zelezniak et al., 2015). As an example, *Bacteroides ovatus* engages in a reciprocal relationship with other bacteria in the gut, by degrading complex polysaccharides, not for its own use, but to support the partner bacteria from which it derives benefit (Rakoff-Nahoum et al., 2016).

Further analysis of the co-occurrence networks can reveal important insights into the patterns of microbial community organization. The analytical methods come from network inference techniques originally developed in computer science, where the "objects" (e.g., the microbial taxa in figure 6.4B) are termed nodes, and the connections are called edges. One of the most consistent results to emerge from network analysis of microbial communities, whether associated with animals or other habitats, is that the connections between the nodes are not randomly distributed. In many instances, the network has a power law distribution (also known as a scale-free distribution), meaning that a few nodes are very strongly connected and many are weakly connected. The highly connected nodes are called hubs, and they are of great interest because they are predicted to include ecologically important taxa, especially keystone species, i.e., species with a disproportionately large effect relative to their abundance. However, as in other aspects of co-occurrence networks, we should be circumspect in interpreting these patterns. In particular, we cannot equate hubs and keystone taxa because: first, some hubs are simply taxa with strong niche overlap with other taxa; and, second, some keystone species do not have many direct connections but exert their effects by indirect interactions that may cascade through the network (Berry and Widder, 2014).

In summary, network analyses and related approaches have great potential to describe the patterns of microbial community structure and identify possible underlying processes including likely candidate taxa and interactions that may be critically important for the overall function of the community. However, because network associations can have multiple ecological explanations (see above), any interpretation should always be validated experimentally. We can anticipate great advances in this area in the coming years, including a broadening of the focus to include cross-kingdom interactions and the trophic structure of the communities. Understanding of both these topics is developing rapidly, including bacterial-protist interactions (chapter 2, section 2.4.1) and the roles of predation and parasitism in animal-associated microbial communities discussed below (section 6.3.3), but, to date, these interactions have barely been considered from a community perspective.

6.3.3. PREDATORS AND PARASITES IN ANIMAL MICROBIOMES

In ecological terms, predators and parasites have the common feature that both engage in +/− interactions (one organism benefits at the expense of the other). Predation and parasitism can play a decisive role in shaping some communities of macroorganisms. Where predators consume abundant prey, competitive interactions among the organisms in the lower trophic level(s) can be reduced, resulting in increased diversity, and parasites of animals and plants can, similarly, have major effects on competitive and other interactions among their hosts and between host and nonhost taxa, with likely ecosystem consequences (Hudson et al., 2006). However, we should not assume that communities of macroorganisms and microorganisms are necessarily structured in the same ways. In particular, anoxic habitats tend to support very "flat" ecosystems because of energy limitations: although some anaerobic communities include predators, such as bacterivorous protists, secondary consumers are rare and higher trophic levels are unknown (Fenchel and Finlay, 1995). We should, therefore, predict that the predominantly anaerobic microbiome of the gut in large animals is limited, at most, to just two trophic levels, but that the structure of microbiomes in some oxic animal habitats may be more complex. At present, we lack the data needed to test this generality, but tantalizing evidence is available on two very different groups: predatory microorganisms and bacteriophages.

Predatory bacteria are detected among the low abundance taxa in many 16S amplicon inventories of animal microbiomes. One distinctive group of predatory bacteria is the *Bdellovibrio*-and-like organisms (BALOs), which are members of the δ-proteobacteria that actively hunt other Gram-negative bacteria. *Bdellovibrio* invades the periplasm (the space between the two membranes) of its prey, where it multiplies, killing the bacterial cell. Because they are obligate aerobes, BALOs are most likely to have community-level effects on microbiomes in oxic conditions. Intriguingly, the BALO *Halobacteriovorax* has high prevalence in the microbiome of corals bearing photosynthetic (hence oxygen-producing) symbionts, and a co-occurrence analysis of the bacterial communities in corals yielded highly significant correlations between *Halobacteriovorax* and various other bacteria, including potential prey (Welsh et al., 2016). Candidate predatory taxa in the microbiomes in anaerobic habitats, e.g., mammalian guts, include various phagotrophic protists, especially ciliates and flagellates. Members of these protist groups are known to play important roles in the degradation of complex dietary polysaccharides (for example in termites and ruminant mammals), but the incidence and ecological significance of predation is largely unknown.

By contrast to the relatively low abundance populations of predatory taxa, phages are very abundant in many habitats. In particular, phage particles become trapped in animal mucus that, for example, lines the gut epithelium of mammals and many other animals and protects the surface of various aquatic animals (e.g., corals, many worms), and it has been proposed the protein motifs contributing to the capsids of some phages may bind specifically to mucin glycoproteins (Barr et al., 2013). In gnotobiotic mice colonized with a select set of human-derived bacteria, orally administered phage caused transient changes in relative bacterial abundance (Reyes et al., 2013), but whether these effects translate into regulation of community composition in natural communities is uncertain.

Lytic phages can, however, promote the populations of their bacterial "prey" in certain ecological circumstances (Obeng et al., 2016). In particular, lytic phages promote the growth and stability of biofilms produced by various bacteria, including *E. coli*, *Staphylococcus aureus*, and *Shewanella*, probably because the lysed cells release DNA, which stabilizes the biofilm, and nutrients that support growth of neighboring bacterial cells. Additionally, phage-mediated lysis can be required for some (but not all) bacteria to release toxins, such as certain bacteriocins, important in competitive interactions. Where the lysed cell is a member of a clonal population that benefits from the lysate, these phage-bacterial interactions are likely to be mutualistic.

6.3.4. HOST DETERMINANTS OF MICROBIAL COMMUNITY COMPOSITION

Host traits are predicted to have a defining effect on microbial community composition, simply because the host is the environment within which the microbiome functions (figure 6.1). As for abiotic environments, an animal is an environmental filter: some microorganisms can survive and proliferate under the conditions and resources of the host environment, and others cannot. An example of highly restrictive conditions imposed by the animal host is the very acidic environment in the stomach of vertebrates and equivalent gut region of some invertebrates. For example, the acidic region of the midgut of *Drosophila* fruit flies has pH 2, and evidence that it plays an important role in controlling microbial colonization is illustrated by studies of *Drosophila* genetically modified to reduce the acidity of this region to ca. pH 4 (Overend et al., 2016). Specifically, key members of the gut microbiota of healthy *Drosophila* had significantly elevated abundance in the *Drosophila* with the acidic region at pH 4, relative to the controls at pH 2 (figure 6.5). The acidity of the human stomach is also a major determinant of the abundance and composition of the stomach microbiota. Important evidence comes from the use of proton pump inhibitors (PPIs). As well as mediating suppression of gastric acid secretion in patients with esophagitis and

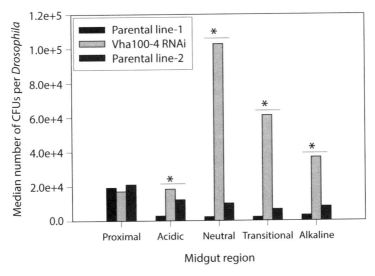

FIG. 6.5. Acidic region of the animal gut as an environmental filter. Increased abundance of gut bacterium *Acetobacter tropicalis* in *Drosophila* with expression of the H+-ATPase gene *vha110–4* suppressed in the acidic region of the gut by RNA-interference (Vha100–4 RNAi) relative to parental controls. (Redrawn from Fig. 7 of Overend et al. [2016].)

gastroesophageal reflux disease, PPIs can cause bacterial overgrowth and changes to the relative abundance of bacteria in both gastric fluid and the esophagus (Amir et al., 2014). These effects may have biomedical importance because it has been suggested that PPI-induced changes in the microbiota may be linked to one type of esophageal cancer (esophageal adenocarcinoma), but causality has not been established definitively.

As well as being the habitat in which microbial partners live, the animal host is an active participant in interactions involving the microorganisms. In particular, the animal immune system plays a central role in shaping the composition and abundance of the microbiota (chapter 4). From an eco-logical perspective, the immune system has many parallels with a predator. Just as choices made by a predator can influence the relative abundance of different prey items, the interactions between the immune system and associated microorganisms can determine the composition of both parasites and symbionts (section 4.4.4). Interestingly, ecological modeling based on Lotka-Volterra models with multiple "prey" (microbes) and "predators" (dif-ferent components of the host immune system) yield a diversity of dynamics, depending on the pattern of interactions between "prey," the quantitative relationship between the immune response and abundance of the microbial partners, and time delays in mounting an immune response (Fenton and Perkins, 2010). These approaches are starting to be applied to microbiome studies, e.g., Rolig et al. (2015), and, as with similar approaches focusing on among-microbe interactions discussed below (see section 6.4.2), they have great potential to illuminate the essential host-microbiota interactions that define microbial abundance and composition.

6.3.5. THE RELATIVE IMPORTANCE OF AMONG-MICROBE INTERACTIONS AND HOST-MICROBE INTERACTIONS

We now have evidence for a role of many positive and negative interac-tions involving different microbes and host in shaping the abundance and composition of the animal-associated microbial communities. The next step is to "put this together," to understand the relative importance of the different drivers and how they interact. Are particular interactions of pre-eminent importance, and does the relative importance of different types of interaction vary with the developmental stage and health status of the host and with environmental circumstance? Resolving these questions is a high priority for the field. This information will ensure that research effort and biomedical strategies for human health are focused on the most important interactions. It has been argued that, if interactions are driven predominantly by the host, then the therapeutic focus should be on the host, requiring procedures that are tailored to host genotype or health status, but if they are microbe-driven, then the focus should be placed firmly on the microbial interactions (Faust and Raes, 2016).

6.4. Functions of the Inner Ecosystem

6.4.1. ECOSYSTEM PROPERTIES OF A SIMPLE ANIMAL MICROBIOME

There is an increasing interest in applying the concepts of ecosystem science to understand and predict the impact of the microbiome on host health and fitness. Just as a tropical rainforest functions in nutrient cycling, protection against soil erosion, and the regulation of climate, the microbiome of an animal has multiple functions (e.g., provisioning of nutrients, protection against pathogens) that shape the health and fitness of the animal host. Key elements of ecosystem function relate to stability and diversity, as summarized in figure 6.6.

The greatest value of ecosystem approaches to microbiome science will, undoubtedly, relate to complex microbial communities associated with animals. Nevertheless, these approaches have already made an important contribution to understanding how relatively simple microbial communities respond to environmental perturbation. This is illustrated by research conducted over many years on the association between shallow-water corals and their algal symbiont, the dinoflagellate *Symbiodinium*. From the perspective of the inner ecosystem (i.e., an animal and its microbiome), *Symbiodinium* algae display the ecosystem functions of primary production, nutrient cycling, and promotion of host calcification (skeleton production). A major natural perturbation to this association is elevated sea water temperatures, which can trigger the expulsion of algal symbionts, leaving the white coral skeleton visible through the animal tissues. Coral bleaching causes reduced growth and reproductive rates of corals, as well as increased susceptibility to disease and elevated mortality rates (Brown, 1997; Douglas, 2003). Many corals, however, recover from a bleaching event, i.e., they are resilient to the perturbation (figure 6.6A). *Symbiodinium* genotype is a major determinant of the bleaching response of corals. Specifically, *Symbiodinium* can be assigned to phylotypes (A, B, C, etc.), including many bleaching-susceptible genotypes in phylotype C and bleaching-resistant genotypes in A, B, and D. This experimentally verified variation in bleaching susceptibility is evident in the bleaching patterns of natural populations of corals. For example, elevated sea water temperatures cause highly localized bleaching in the Caribbean coral *Montastraea annularis*. Shaded portions of the coral colonies containing phylotype C bleach readily, while the exposed colony tops with phylotype B are bleaching-resistant (Rowan et al., 1997).

Coral colonies that have recovered from a bleaching episode are often more resistant to subsequent episodes of high temperature. In some situations, this change can be explained by physiological acclimation of the coral tissues (Brown et al., 2000) but, in many coral species, it is associated with a shift in the composition to an alternative stable state (figure 6.6B). This is elegantly illustrated by research on the Pacific coral *Acropora millepora*, which bears the bleaching-susceptible phylotype C in the cooler, more southerly waters off Australia while the bleaching-resistant phylotype D is dominant in the warmer northern sites (Berkelmans and van Oppen, 2006). When fragments of coral, known as nubbins, containing phylotype

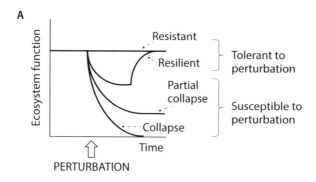

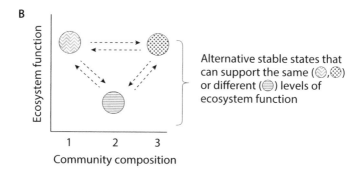

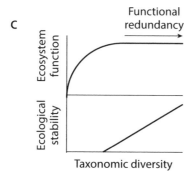

C were transferred from two cool sites (Keppel Island and Davies Reef) to a warm site, Magnetic Island, they displayed some bleaching, but most of the transplanted nubbins survived and recovered. However, the inner ecosystem within nubbins from one site (Davies Reef) was unchanged, recovering the original *Symbiodinium* phylotype C2*, while the nubbins from North Keppel Island had shifted to the alternative state of phylotype D. The replacement algae of phylotype D were, almost certainly, present in the pretransplant corals but in very low numbers (Bay et al., 2016). When the nubbins were tested for bleaching susceptibility (figure 6.7), only the transplanted coral nubbins that had switched *Symbiodinium* phylotype displayed resistance to bleaching, at least to 31°C.

These studies of the coral-*Symbiodinium* associations demonstrate how composition of the microbiota can influence ecosystem function (figure 6.6A), and how environmental perturbations can trigger shifts in composition to different alternative states (figure 6.6B). How do complex microbial communities in animals respond to perturbation? Do they shift between alternative stable states, and are these states advantageous to the animal host? In addition to these questions, there is the issue of diversity. Do microbiomes conform to the prediction from ecosystem science (figure 6.6C) that microbial diversity enhances stability of ecosystem functions subject to perturbation?

FIG. 6.6. (Opposite page) Ecosystem functions. A. Ecosystem stability, i.e., low variability over time despite perturbation. A stable system can be resistant to the perturbation (no effect on indices of ecosystem function) or resilient to the perturbation, i.e., recover to the original state after perturbation, while susceptible ecosystems display reduction or collapse of ecosystem function. B. Alternative stable states. The composition and diversity of ecological communities can be influenced by the order and timing of colonization, a phenomenon known as the priority effect. Alternative stable states (also known, for example, as alternative attractors, multiple stable points) arise under identical environmental conditions as a result of differences in assembly history, and an ecosystem can transition between these different states when perturbed. However, many real communities may be less discrete and less stable than the expectation of alternative stable states. C. Relationship of biodiversity with ecosystem function and stability, as predicted by stability-diversity models. A taxonomically diverse community may include much functional redundancy, enabling system function to persist despite the loss of some taxa mediating a particular function; and a more diverse community is also likely to include taxa with different tolerances to the perturbation, such that the system as a whole can function under a greater range of conditions than is optimal for any one member of the community. High diversity has also been argued to reduce the strength of individual pair-wise interactions among members of the community, with the expectation that the loss of one species would have limited knock-on effects on the abundance and function of other species. The significance of stability, redundancy, diversity and alternative stable states for ecosystem function are reviewed by McCann (2000).

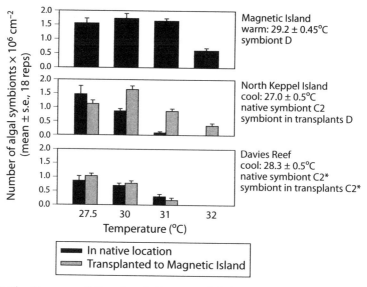

FIG. 6.7. Bleaching susceptibility of corals. Response of algal symbiosis in the Australian coral *Acropora millepora* to elevated temperature (≥30°C). The corals tested were either in their native location or, for the two cool locations (North Keppel Island and Davies Reef), had been transplanted to the warm location (Magnetic Island). The tolerance of elevated temperature in coral transplanted from North Keppel Island to Magnetic Island is correlated with a shift in algal symbiont from phylotype C2 to phylotype D. The summer temperatures (Dec–Feb) at each site are shown. (Redrawn from Fig. 6 of Berkelmans and van Oppen [2006].)

6.4.2. DIVERSITY, STABILITY, AND ALTERNATIVE STABLE STATES

To address how the composition and diversity of the microbiome can influence the function of animal ecosystems that include many different microorganisms, we first need to consider a technical issue. The reported α- and β-diversity (figure 6.3) of animal microbiomes is strongly influenced by sampling effort and taxonomic resolution. Relatively shallow sequencing (such that some low-abundance taxa are undetected) and high taxonomic resolution will enhance reported β-diversity, both among host individuals and over time within a single host individual, while deep sequencing and low taxonomic resolution will emphasize among-sample similarity, including the apparent detection of a core microbiome (the concept of the core microbiome is addressed in chapter 3, section 3.2.2). As a result, studies with different sampling strategies can yield rather divergent conclusions. For example, linked to differences in sampling design, some studies on the temporal variation in the gut microbiota of humans emphasize considerable variability, while others note stability over time (Caporaso et al., 2011; David, Maurice, et al., 2014; Faith et al., 2013; Oh et al., 2016).

Despite these differences, these time series studies provide insights into the impact of different environmental factors in daily life on microbiota composition. For example, in the year-long study of David, Materna, et al. (2014), no consistent effects of many day-to-day variations, including diet, mood, or exercise, on the microbiota were evident, but foreign travel had a large but transient effect and *Salmonella* food poisoning caused persistent changes that included the apparent extinction of ca. 15% of the operational taxonomic units (OTUs) in preinfection samples. These results suggest that the gut microbiota varies in its resilience to different types of perturbation.

These studies on the temporal variation in the composition of the gut microbiota provide valuable insights, but they were not designed to consider either the effect of microbiota variation on host indices of ecosystem function or whether any of the variation could be considered as alternative communities. There are, however, some tantalizing indications that the microbiome in humans can adopt alternative states with potentially profound effects on human health. One important study (M.I. Smith et al., 2013) focused on the relationship between the gut microbiota of very young children (<3 years) in Malawi and the incidence of kwashiorkor, a debilitating disease caused by inadequate dietary protein, in which children fail to gain weight even when provided with a nutritious diet. Comparison between the gut microbiota metagenomes of "discordant" twins (i.e., one healthy twin and one twin with kwashiorkor) revealed significant functional differences between the two groups (figure 6.8A). Whereas the microbiota in the healthy twins matured over time, the microbiota of the kwashiorkor twins did not vary significantly over time, appearing to be constrained to an alternative state. To investigate how these differences were related to kwashiorkor, microbiota from discordant twins were introduced to germ-free mice. When fed on the standard diet of the children (Malawian diet), the mice with the microbiota from healthy children sustained normal weight, while those with the kwashiorkor microbiota lost 40% of their body weight, indicating that the microbiota played a causal role (figure 6.8B). This difference was associated with low abundance of *Bifidobacteria* and *Lactobacilli*, and elevated levels of *Bilophila wadsworthia* (a sulfite-reducing bacterium related to *Desulfovibrio*) and *Clostridium innocuum*, as well as differences in the amino acid and SCFA titers in the mice. Interestingly, the mice with the kwashiorkor-microbiota gained weight when transferred to a nutritious diet, and this intervention had a protective effect against extreme weight loss when the mice were returned to the Malawian diet.

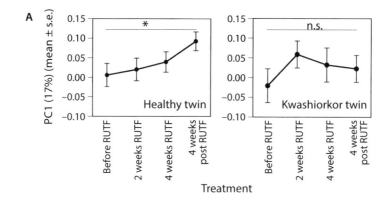

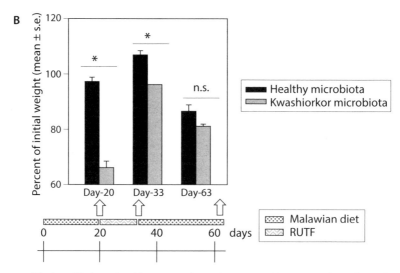

FIG. 6.8. The kwashiorkor microbiome as an alternative state. A. First principal coordinate (PC) axis score of functional traits (KEGG enzymes) of gut microbiome sampled from twin pairs discordant for the nutritional disease, kwashiorkor, showing maturation of the healthy, but not kwashiorkor, twin, over the time of study. RUTF, a peanut-based "ready-to-use therapeutic food." (Fig. 1A of M.I. Smith et al. [2013].) B. Weight of germ-free mice administered the microbiota from healthy or kwashiorkor twins at day 0. (Data from Fig. 2A of M.I. Smith et al. [2013].)

The greatest importance of the study of M.I. Smith et al. (2013) is that it offers a basis for microbial therapy to alleviate kwashiorkor. In addition, this study offers an exemplar of how the microbiome can be "locked" into alternative taxonomic and functional states with very substantial consequences for host health. Alternative, apparently stable states of the human microbiome have been reported for other diseases, including the skin microbiota in relation to ulcer-inducing bacterial infections (van Rensburg et al., 2015) and the gut microbiota in relation to type 2 diabetes

(Qin et al., 2012). There are also reports of distinctively different microbial communities in healthy people. For example, multiple alternative states of the vagina microbiome have been reported, four of which are dominated by different *Lactobacillus* species and a fifth by strictly anaerobic bacteria including the proteobacterium *Prevotella* and Actinobacteria (e.g., *Gardnerella* and *Atopobium*) (Ravel et al., 2011). Longitudinal studies indicated that the microbiota community type was very stable in some women, and switched between types in other women, often over very short time periods, but intermediate community types were very rarely detected (Faust et al., 2015; Ravel et al., 2011). It has, similarly, been proposed that the gut microbiota of adults with a Western lifestyle can be assigned to three types, originally described as enterotypes (Arumugam et al., 2011). The existence of qualitatively different microbial communities in healthy humans has, however, been disputed, with various analyses indicating that these types are better described as ends of a continuum than discrete states (Knights et al., 2014; Koren et al., 2013).

Although the existence and significance of alternative states of the microbiome in healthy people are uncertain, research on the effects of antibiotic treatment provides strong evidence that a microbiome can undergo sharp transitions between alternative states. Oral antibiotics represent a major perturbation for the gut microbiota, causing massive reduction in the abundance of microorganisms and, in some cases, effects that persist beyond the treatment period. In one study on healthy adult volunteers, anaerobic bacteria of the genus *Bacteroides* were greatly depleted by treatment with metronidazole (which specifically targets anaerobic bacteria), and had not returned to the pretreatment abundance two years after treatment (Jakobsson et al., 2010). In a different study, the fluoroquinolone antibiotic, ciprofloxacin, caused dramatic changes to the composition of the microbiota, with the abundance of up to 50% of the taxa affected. The microbiota showed some resilience, tending to return to a composition approaching that of the pretreatment condition within two weeks, although with considerable variation among volunteers (Dethlefsen and Relman, 2011). More detailed information comes from experiments conducted on mice. In particular, a single dose of the antibiotic clindamycin resulted in a 90% reduction in the diversity of bacteria in the cecum (figure 6.9A), with a blooming of a few taxa, notably undefined Enterobacteriaceae and *Enterococcus* (Buffie et al., 2012). The mice with the antibiotic-induced altered communities of low diversity suffered higher morbidity and mortality when challenged with the pathogen *Clostridium difficile*; and the microbiota in antibiotic-treated mice that recovered from the *C. difficile* infection remained at low diversity and were susceptible to further challenge with *C. difficile*. The data conformed to theoretical predictions of alternative stable states driven by antibiotic-induced perturbation

(Stein et al., 2013), particularly antibiotic-mediated promotion of *Enterococcus* (figure 6.9B).

As these examples illustrate, we cannot understand how animals respond to environmental factors without considering the effects on the microbiome—and these effects can be long-lasting, persisting beyond the period of exposure to specific environmental conditions. However, all these considerations have included an important simplification: that the host is functionally invariant over time. The reality is that animals undergo tightly orchestrated changes over multiple temporal scales ranging from developmental changes to circadian and seasonal rhythms. The evidence for this variation and its functional significance are considered next.

6.4.3. PRIMARY SUCCESSION IN THE INNER ECOSYSTEM

For many animals, embryogenesis proceeds in a microbiologically sterile environment, or nearly so. This means that, at hatching or birth, the neonate animal is a pristine habitat, analogous to a fresh lava flow or the substrate under a recently retreated glacier. This has led to the suggestion

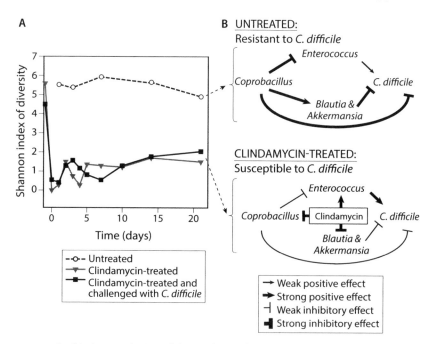

FIG. 6.9. Antibiotic perturbation of the cecal microbiota of mice. A. Reduced diversity of cecal microbiota in mice injected with a single dose of clindamycin (200 μg) on day −1 of the experiment. (Modified from Fig. 4H of Buffie et al. [2012].) B. Predicted interactions among bacteria in the mouse cecum under two alternative stable states, untreated (top) and clindamycin-treated (bottom). (Modified from Fig. 5B of Stein et al. [2013].)

that the colonization history of a developing animal may follow certain predictable patterns that can be described as primary succession, and that deviation from this pattern may be symptomatic or causal of poor host health.

Evidence that the colonization history of neonate animals can follow predictable patterns comes from research on the bacterial communities associated with *Hydra* (Franzenburg, Fraune, et al., 2013). Laboratory populations of *Hydra vulgaris* are colonized by a well-defined microbiota, dominated by the β-proteobacterium *Curvibacter*. The microbiota associated with newly hatched polyps differs from the adult microbiota in two important ways: the diversity is higher (figure 6.10A), and the composition is different, being dominated by *Flavobacterium*. A longitudinal analysis of the microbiota over 14 weeks revealed a reproducible biphasic pattern, involving a transient adult-like community at 2 weeks and then a reversion to high diversity, *Curvibacter*-poor community over several weeks before stabilization of the adult condition (figure 6.10A). In a similar way, the diversity of bacteria associated with the surface of a marine crab *Kiwa puravida* collected from a deep-sea methane vent off Costa Rica declined progressively from eggs through juveniles to adults, and this was coupled with a dramatic change in the composition of the microbiota from domination by γ-proteobacteria (Thiotrichaceae and Methylococcaceae) to a microbiota that was dominated by a *Rimicaris*-like taxon within the Helicobacteraceae (figure 6.10B) (Goffredi et al., 2014). Although these studies did not quantify the among-individual variation and how it varied over developmental age, the data are strongly suggestive of a progressive reduction in both α- and β-diversity as the animals develop (Franzenburg, Fraune, et al., 2013; Goffredi et al., 2014). This pattern is not displayed by all animals. For example, the gut microbiota of the zebrafish varied with developmental age, becoming progressively less like the bacterial communities in the external environment, but this was accompanied by an increase in among-individual variation (i.e., β-diversity) with increasing age of the developing fish (Zac Stephens et al., 2016).

There is a special interest in the pattern of microbial colonization of the human infant because microbial factors in early postnatal development may be predictive, and possibly causal, of health in later life. The first colonizers of the newborn are derived from the fecal, skin, and vaginal microbiota of the mother (Palmer et al., 2007). There have been suggestions that microbes play a role in the development of the healthy human embryo, based on culture-independent detection of bacterial 16S sequences in the placenta and amniotic fluid. However, some purported demonstrations of placental or amniotic microbiotas lack important controls, raising the possibility of artifactual results; and microbial colonization of the placenta and amniotic fluid have been linked to preterm birth (reviewed by Charbonneau et al.,

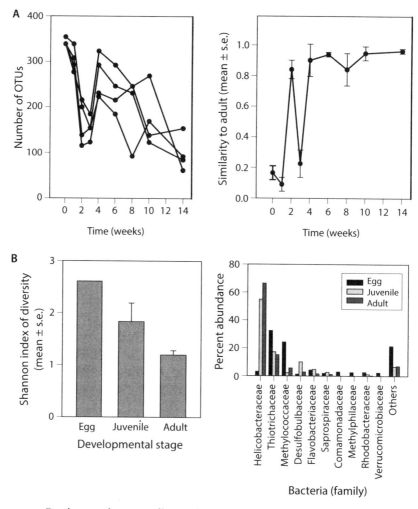

FIG. 6.10. Developmental patterns of bacterial communities in animals. A. Species richness (left) and taxonomic similarity to the adult microbiota (right) in *Hydra vulgaris* over the first 14 weeks after hatching (Redrawn from Fig. 2 and Fig. 4b of Franzenburg, Fraune, et al. [2013]). B. Taxonomic diversity (left) and relative abundance of different taxa (right) at different developmental stages of the crab *Kiwa puravida*. (From data in Table 3 and Table 4 of Goffredi et al. [2014].)

2016), suggesting that the health benefits of microbial colonization are likely to be strictly postnatal.

Multiple studies have demonstrated successional changes to the human gut microbiota over the first months and years of life (Charbonneau et al., 2016; Mueller et al., 2015; Rodriguez et al., 2015). In the healthy baby, the first colonists are commonly replaced by bacteria, especially *Bifidobacteria*, that are adapted to the diet of human milk, and these communities are subsequently reorganized at weaning, with essentially adult-like communities in place at ca. 3 years. A very important driver of the gut microbiota in the first weeks of life is the capacity to utilize the complex human milk oligosaccharides (HMOs), as well as glyco-proteins and glycolipids (Smilowitz et al., 2014). These sugars are 3–20 monosaccharide units comprising a lactose core with glycosidic links to different sugars, including glucose, galactose, N-acetyl-glucosamine, and fucose. Importantly, HMOs cannot be degraded by digestive enzymes of the baby, but can be utilized by *Bifidobacteria* and other beneficial bacteria. Together with other constituents, including immunoglobulins (see chapter 4, section 4.3.3), the HMOs in human milk are believed to control the microbial community in the gut prior to weaning. Indeed, the diversity and complexity of HMOs may promote a high diversity microbial community comprising multiple strains with different metabolic capabilities, none of which can attain dominance that would be deleterious to the host. In other words, the composition of milk has, most probably, coevolved with functional traits of colonizing microbes that are beneficial for the health of the baby. However, many questions remain unresolved. It is not understood how different components of milk contribute to the overall pattern of microbial colonization and its variation over time, nor how variation in the composition of the milk among different mothers may affect the health of the baby. In addition, it is still unclear how, and how much, the composition of the microbiota in the first weeks and months of life influences the composition of the adult-like microbiota in the fully weaned child and, ultimately, the microbiota in the adult.

6.4.4. HOST RHYTHMS AND FUNCTION OF THE INNER ECOSYSTEM

Decades of research have identified the molecular and biochemical mechanisms by which many animals orchestrate their life histories and phenotypic traits to predictable environmental changes, especially season and time of day. However, recent data suggest that the animal rhythms cannot be explained entirely in terms of animal processes, and may be shaped by the inner ecosystem of animal-microbial interactions (figure 6.11A). There are three layers to the response of the inner ecosystem to season and time of day: Does the microbiota respond to the biological rhythms of the host? Does the response of the microbiota reinforce the adaptive rhythmic changes of the host? And does the microbiota modulate the host regulatory networks that orchestrate host rhythms?

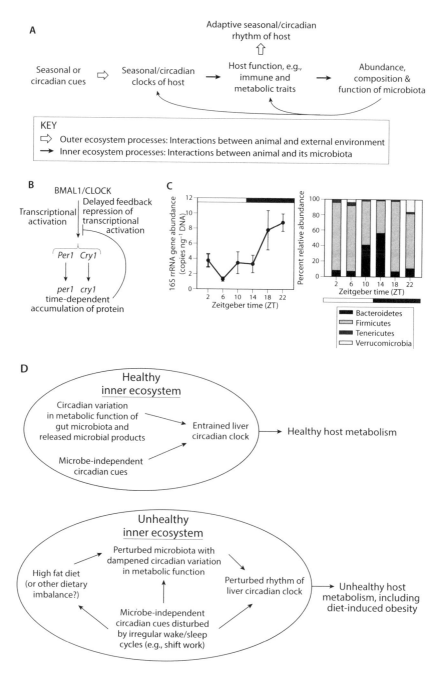

FIG. 6.11. Biological rhythms. A. Biological rhythms driven by interactions between inner eco-system and outer ecosystem processes. B. The core of the mammalian circadian clock is an oscillating transcriptional feedback loop, in which a cryptochrome (CRY1) and the protein Period-1 (PER1) block the activation of their own gene expression by two transcriptional factors,

Evidence that seasonal rhythm of an animal can be underpinned by bi-directional interactions between the animal host and its microbiota comes from research on the brown bear *Ursus arctos* (Sommer et al., 2016), which hibernates in the winter. Metabolic function, especially energy harvesting, differs between the bears in summer and winter: the bears deposit large amounts of fat in the summer, and have elevated serum levels of triglycerides and cholesterol in the winter. A first indication that the gut microbiota participates in this response came from evidence that the microbial composition (as sampled from fecal content) between the bears differed between summer and winter, with high levels of Firmicutes and Actinobacteria in the summer and more Bacteroidetes in the winter. To test whether these differences influence the biological response of the bears to season, the "summer microbiota" and "winter microbiota" were transferred to germ-free mice. Many of the metabolic differences between the bears in the summer and winter were replicated in the mice under uniform conditions, indicating that the microbiota can contribute to the seasonal differences in the metabolic function of the host. This study demonstrates that the response of the brown bear to season is a function of the inner ecosystem, i.e., the microbiota and the animal.

How does the microbiota influence the biological rhythms of the animal host? Unique insight comes from research on the circadian rhythm in the association between the squid *Euprymna scolopes* and luminescent bacteria *Vibrio fischeri* that inhabit the squid light organ. The bacteria luminesce at night, providing camouflage for the squid by obscuring the dark silhouette of the squid against down-welling light from the moon and stars. The squid has two genes for cryptochromes (*escry1* and *escry2*), which are expressed in the head and light organ. In other animals, these light-sensitive proteins regulate the circadian clock and are much more strongly expressed during the day than the night. In the squid light organ, however, expression of *escry1* is elevated during the night, when the bacteria are luminescent, and this pattern of expression required both bacterial light and microbial cell wall components (Heath-Heckman et al., 2013). These data suggest that the bacterial symbionts regulate a key component of the molecular machinery controlling the circadian clock in the squid light organ.

Developing research on the circadian rhythm in mammals suggests that the interaction between *Vibrio* and cryptochrome gene expression in the squid light organ has parallels in other animal-microbial systems. This field of inquiry is still in its infancy and our understanding is fragmentary, but an

BMAL1 and CLOCK. C. Circadian variation in the abundance (left) and composition (right) of bacteria in the cecum of mice fed on the healthy diet of mouse chow. (From Fig. 2D and Fig. S2 of Leone et al. [2015].) D. Candidate role of the gut microbiota in entrainment of the liver circadian clock for healthy metabolism (top) and diet-induced obesity (bottom).

indication of its importance comes from a study of the circadian clock in the liver of the mouse (Leone et al., 2015). The diurnal variation in expression of the mouse cryptochrome gene *cry1* and the other three genes that comprise the core of the clock (figure 6.11B) is disrupted in germ-free mice, suggesting that the gut microbiota may be important in entraining the liver circadian clock. This interpretation is reasonable because, firstly, the abundance and composition of the bacterial community in the mouse cecum varies with time of day (figure 6.11C); and, secondly, the diurnal variation in the microbiota is severely dampened in mice on a high fat diet, with correlated perturbation in the liver circadian clock (Leone et al., 2015). Together, these data provide indications that the gut microbiota may communicate with the liver clock via circadian variation in the production of SCFAs and possibly other metabolites. These results are congruent with the growing perspective that human metabolic diseases, including obesity and type 2 diabetes, may be causally linked to the interactive effects of disrupted circadian cues (for example, by shift work), inappropriate diet, and perturbations to the microbiota (figure 6.11D).

In summary, there is now a growing body of data suggesting that the biological rhythms of animals that have evolved in response to highly predictable fluctuations in the outer ecosystem are mediated by complex interactions in the inner ecosystem of animal and microbial cells. This should not be so very surprising, since the cellular clocks of animals are one of many regulatory circuits that, as discussed in chapter 2, evolved and diversified in the context of ancient interactions with resident microbial communities.

6.5. Summary

An important goal of microbiome science is to explain and predict the interactions between an animal and its microbiota, including how these interactions respond to change in environmental circumstance and host condition (e.g., developmental stage, health status). Because host-microbial interactions are inescapably complex, ecological principles provide a powerful framework to achieve this goal.

From the perspective of the individual animal, the animal and its microbiota comprise an inner ecosystem, nested within and interacting with the outer ecosystem of external organisms and the physical environment. The perspective of the microbial partners is different: because each animal has a finite size and lifespan, the persistence of the microbial partners depends on continual flux between animals, often via the external environment. The extent to which animal-associated microorganisms sustain free-living populations varies widely, from taxa that are unknown apart from the host, through taxa with transient free-living populations to self-sustaining populations in the external environment (section 6.2).

Returning our focus to the inner ecosystem within an individual animal, the composition of the microbiota is determined by interactions among microorganisms, as well as between microorganisms and the host (section 6.3). Valuable information can be gained from the pattern of co-occurrence (or coexclusion) of microorganisms across different host individuals, together with experimental analysis, often using simplified microbial systems, to discriminate among multiple alternative processes that can underlie the observed patterns. For example, co-occurrence may signify mutually beneficial interactions or niche overlap. Generally, negative correlations dominate the co-occurrence networks in animal-associated microbial communities. It has been suggested that at least a fraction of these negative correlations represent antagonistic interactions that promote community stability, while cooperative interactions can cause instability by promoting coupled population increase or loss of mutually dependent taxa. There are also indications that these symmetrical $+/+$ and $-/-$ interactions among bacteria likely interact with $+/-$ interactions with predators (e.g., predatory bacteria, bacterivorous protists) and phage, and also with the host. The relative importance of among-microbe and host-microbe interactions in shaping the microbiota community composition and function is poorly known, and an important issue for future research.

Animal health and fitness are promoted by stability in the taxonomic and functional composition of the microbiota (section 6.4). It has been suggested that the inner ecosystem can adopt different, apparently stable states that are correlated with (and, in some cases, predictive of) host health. These states can be induced by environmental perturbation, such as antibiotic treatment or disease. Alternative states have also been proposed in different healthy hosts (e.g., for the gut microbiota of different people), although some analyses are indicative of continuous variation of the microbiome among healthy individuals. Microbial communities also undergo broadly predictable changes in taxonomic and functional composition associated with animal development, comparable to successional changes in other ecosystems, as well as changes that are correlated with fundamental biological rhythms linked to time of day and season. Initial data suggest that the animal responses to these external environmental factors are reinforced and stabilized by interactions with the microbiota of the inner ecosystem.

7

Evolutionary Processes and Consequences

7.1. Introduction

The health and fitness of humans and other animals are founded on their association with microorganisms. This statement is based on a substantial and growing body of evidence and is widely accepted today. However, it would have posed major problems for most biologists through much of the twentieth century (Sapp, 1994). The difficulty lay at the intersection between ecology and evolutionary biology. The Lotka-Volterra equations of population increase provided a firm theoretical basis for the verbal argument that antagonistic interactions tend to stabilize populations of interacting organisms via negative feedback loops, while cooperative interactions lead to uncontrolled mutual increase and should, consequently, be rare and transient. In parallel, theoretical and empirical studies demonstrated that selection favors traits advantageous to the individual relative to traits that are good for the group or species. Altruism was explained with great elegance in terms of kin selection, at least in insect societies, and mutualisms and cooperation were regarded as curiosities of nature, of little general significance.

It is now realized that, although the preceding paragraph is partly correct, it is incomplete. A firm basis of theory has developed to explain the evolution and persistence of cooperation, including between unrelated partners (the focus of microbiome science), and the theory has been tested and developed by experimental studies (Dimitriu et al., 2014; Nowak, 2006; Oliveira et al., 2014; Rainey and Rainey, 2003; West et al., 2011). These developments demonstrate that the standing principles of evolution by natural selection can accommodate microbiome science and that, contrary to some claims, a new theory of biology is not required.

In this way, microbiome science is supported by a mature literature on biological interactions and their evolutionary consequences. Combined with ever-increasing sequencing capabilities that enable study of the phylogenetic relationships among microorganisms and their functional capabilities, these advances are leading to new insights into the selection pressures that bring partners together, shape their evolutionary trajectories, and, ultimately, influence organismal diversity. In this chapter, these three topics are considered in turn, with reference to both animal associations with complex communities and associations with single microbial partners, especially intracellular symbionts. Some experimental approaches are currently most feasible for one-host-one-symbiont relationships, yielding general principles that can be applied to more complex interactions between communities of microorganisms and the host.

Section 7.2 addresses the fitness consequences for partners in animal-microbial associations, and the underlying interactions. Reciprocal exchange of services and byproduct mutualism play key roles in maintaining associations, while some interactions can best be understood as exploitation or addiction. Section 7.3 focuses on microbial partners specialized to specific host lineages, such that the phylogeny of the microbial partners maps onto that of their hosts. There is now excellent evidence for coevolutionary interactions in some associations with congruent phylogenies, and the incidence and distribution of coevolution in animal-microbial associations are discussed. Section 7.4 takes a broader perspective, focusing on the growing evidence that animal-microbial associations can increase the rates of evolutionary diversification of the partners. We finish with a topic of considerable uncertainty: whether microbial partners promote speciation of their animal hosts by processes that are different from symbiosis-independent speciation.

7.2. Costs and Benefits

7.2.1. RECIPROCITY

Cooperation between unrelated organisms is most readily explained in terms of reciprocal exchange of services. Reciprocity is driven by the balance sheet of costs and benefits. Cooperative associations evolve and persist where partners provide each other with complementary services, and the benefit of the service received from the partner is greater than the cost of producing the reciprocal service (figure 7.1A). Reciprocal exchange is the basis of many cooperative relationships in biology, as well as in human society (Taborsky, 2013). Reciprocity is most easy to investigate where the services are functionally similar, e.g., exchange of nutrients, but can be extended to different types of services, e.g., one organism protects its partner from natural enemies in exchange for nutrients received from the partner.

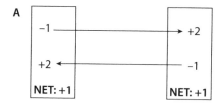

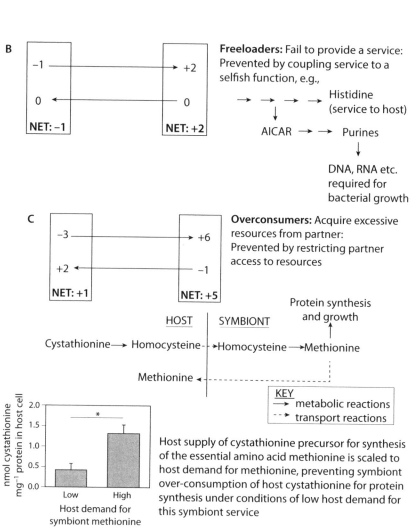

FIG. 7.1. Reciprocity in animal-microbial associations. A. Mutual benefit through reciprocal exchange of costly services. B. Exploitation by a freeloader (also known as the tax dodger). C. Exploitation by an over-consumer (also known as tragedy of the commons). The routes by which the host gains protection from a freeloader and overconsumer are illustrated by host control of the release of two essential amino acids, histidine and methionine, by the symbiotic bacterium *Buchnera* in aphids (see text for details). (B is redrawn from Fig. 3b of Thomas et al. [2009], and C is redrawn from Fig. 3c of Russell et al. [2014].)

In many animal-microbial associations, the microbial services can be defined in terms of metabolic capabilities (see chapter 1, Table 1.1A). Because the lineage giving rise to animals was metabolically impoverished in comparison to many microorganisms, especially bacteria, many animals benefit by gaining access to the products of microbial metabolism. As described in chapter 1, animals variously derive: supplementary carbon, for example from intracellular photosynthetic algae (e.g., in corals) and from gut microorganisms that degrade complex polysaccharides (e.g., in termites); essential nutrients, such as B vitamins (e.g., in insects that feed through the life cycle on vertebrate blood); and bioactive compounds that confer protection against natural enemies (e.g., bacterial polyketides in benthic marine animals). These services can, in many systems, be quantified as the synthesis and release rates of specific metabolites, and the concentration of these products in the tissues or body fluids of the animal partner.

The reciprocal services provided by the animal to its microbial partners are usually described in terms of the animal as a habitat (Table 1.1B). Animals are nutrient-rich, in comparison to many abiotic habitats; they are well-defended by their immune system against the depredations of saprophytic microorganisms; and they are water-rich relative to most terrestrial habitats. Furthermore, because most animals function at larger spatial scales than most microorganisms, they can promote microbial dispersal. Although it is intuitive that these services would confer fitness advantage for microorganisms, these benefits are generally difficult to quantify. Thus, the photosynthetic compounds released by the algal population in a coral or the fermentation products released by the community of anaerobic microorganisms in the cecum of a mouse can, in principle, be quantified; and the contribution of these products to the energy budget and fitness of the coral or mouse can be determined. But the reciprocal services provided by the host have rarely been quantified and, very often, it is uncertain how these services should be quantified. There is no simple answer to such simple questions as: Do microorganisms have enhanced access to nutrients in the host relative to the free-living environment, and greater protection from unfavorable abiotic conditions and natural enemies in the host than in the free-living environment? Consequently, it is widely assumed that animal-microbial interactions are reciprocal relationships but this has rarely been demonstrated quantitatively. I return to this issue in section 7.2.2.

There is one further issue to address relating to reciprocity: conditions where providing a service is cost-free, or nearly so. Cost-free services are often known as byproduct mutualism. The examples I provided in the previous paragraph include one costly service and one apparently cost-free service. The photosynthesis-derived sugars and related compounds released from algal cells in corals accounts for 30–90% of the fixed carbon, representing a substantial diversion of resources to support the energy metabolism of the host (Tremblay et al., 2012). In contrast, the release of short chain fatty acids (SCFAs) and other fermentation products from the microorganisms in the mouse cecum is, apparently, cost-free. As discussed in chapter 5 (section 5.4.1), these compounds are waste products of the anaerobic metabolism, and their value to the host is incidental to the microbial partners. Another widespread example of a cost-free service is host production of waste nitrogenous compounds that are utilized as nitrogen sources by microbial partners. Scenarios can be envisaged where the composition or amount of fermentation products, nitrogenous waste products. etc. is modified from the metabolic optimum for the producer, in response to selection exerted by the partner. For example, the production of compounds that the partner cannot metabolize may be minimized and, as discussed in chapter 5 (section 5.4.2), the production of metabolically costly compounds may be increased as an honest signal of good intent. Further research is required to investigate the evolutionary lability of cost of services, including the incidence of evolutionary transitions between cost-free and costly services, in animal-microbial associations.

7.2.2. EXPLOITATION

Reciprocal interactions are open to exploitation by cheaters (Ghoul et al., 2014). Cheating can take two alternative forms. There are freeloaders that fail to provide any service but continue to benefit from services provided by the partner (figure 7.1B), and there are overconsumers that abstract excessive resources from the partner (figure 7.1C). Freeloaders in biological systems are akin to tax dodgers in human societies, and overconsumers are often compared to the tragedy of the commons, where prudent use of common grazing grounds in Medieval England was reportedly undermined by individuals who put excessive numbers of animals onto the commons for their individual economic benefit. I understand from historians of Medieval England that strong social controls generally prevented selfish overgrazing. As considered below, analogous controls have evolved in animal-microbial symbioses.

Various routes preventing exploitation by microbial partners have evolved in animal hosts. Most of this research has focused on one-host-one-symbiont associations. Figure 7.1 provides illustrative examples of host mechanisms that counter microbial cheating in one animal association, the symbiosis between the pea aphid *Acyrthosiphon pisum* and its intracellular γ-proteobacterium *Buchnera aphidicola*. Aphids feed on plant phloem sap, a diet deficient in the 10 essential amino acids (the amino acids that contribute to protein but cannot be synthesized by animals), and these insects gain supplementary essential amino acids from their *Buchnera* bacteria. The released amino acids account for 30–50% of the total synthesized by the bacterial cells, representing a substantial metabolic cost that is interpreted to limit protein synthesis and growth of the *Buchnera* cells. With respect to one amino acid, histidine, freeloading (i.e., releasing less histidine) is prevented by coupling histidine synthesis to the synthesis of compounds required for *Buchnera* growth (Thomas et al., 2009). Specifically, a byproduct of one reaction in histidine synthesis, AICAR (5-aminoimidazole-4-carboxamide ribonucleotide), is an intermediate in the synthesis of purines, which are essential building blocks for DNA, RNA, ATP, etc. In many bacteria, AICAR from the histidine biosynthetic pathway makes a relatively small contribution to purine synthesis because metabolic flux through the purine biosynthetic pathway is far higher than through the histidine synthesis pathway. In *Buchnera*, however, the purine biosynthetic pathway is proximally truncated, such that AICAR is the precursor and sourced entirely from the histidine biosynthetic pathway (figure 7.1B). This means that metabolic flux to histidine biosynthesis must exceed the bacterium's requirements, in order to meet its requirements for purines. (Accumulation of metabolic intermediates in histidine synthesis would likely be toxic.) In this way, *Buchnera* has to overproduce histidine to sustain its own growth. It is compelled by the structure of the metabolic network (particularly the proximal truncation of the purine biosynthetic pathway) to provide the service of histidine release, thereby precluding freeloader traits.

In principle, the *Buchnera* symbionts of aphids could also cheat by overconsuming host-derived precursors of essential amino acids, thereby achieving excessive rates of protein synthesis and growth, as well as meeting the host requirement for essential amino acids (figure 7.1C). With respect to the essential amino acid methionine, overconsumption is constrained by tight host controls over supply of the precursor. Thus, the concentration of the cystathionine precursor of methionine synthesis varies according to the host demand for methionine, where high demand is imposed by rearing the aphid on a methionine-free diet and low demand on a diet containing methionine (Russell et al., 2014). Although it is not currently known how the host scales the availability of cystathionine to methionine demand, nor whether synthesis of other essential amino acids is regulated by precursor supply, this system provides an exemplar of how hosts can protect against overconsuming cheaters.

Of course, the animal host can also cheat on its symbionts, either by failing to provide services or abstracting excessive resources from the symbionts. Host exploitation of its microbial partners has proven to be far more difficult to investigate than the reverse largely because, as discussed in section 7.2.1, the services provided by the host tend to be difficult to quantify. Indications that the host may be exploitative come from an intriguing study conducted not on an animal, but on the freshwater ciliate *Paramecium bursaria.* This single-celled protist is densely populated by cells of the green alga *Chlorella*, which are photosynthetically active, releasing much of the photosynthetic carbon in the form of the disaccharide maltose (figure 7.2A). Lowe et al. (2016) compared the growth rates of both the *Paramecium* and the *Chlorella* in the symbiosis and in isolation, and interpreted higher growth rates in symbiosis than in isolation as benefit, and lower growth rates as cost (figure 7.2B). The significance of the symbiosis for the *Paramecium* host varied with irradiance. In darkness, maintaining the *Chlorella* was costly, presumably because the algal cells consumed host resources without any reciprocated service (i.e., the *Chlorella* were freeloaders). However, as the irradiance increased, so did the benefit to the *Paramecium*. The impact of irradiance on the growth rates of the *Chlorella* were very different. Growth rates increased more slowly with irradiance for the *Chlorella* cells in symbiosis than in culture, and even declined at the highest irradiance tested, indicating that it is costly for the *Chlorella* to be associated with *Paramecium*. Further study of the photosynthetic properties of the *Chlorella* in symbiosis and isolation led Lowe et al. (2016) to conclude that the symbiotic *Chlorella* were nutrient-limited at high irradiances, suggesting that the host was acting as a freeloader, i.e., failing to provide resources. In terms of growth rates, the association between *Paramecium* and *Chlorella* is exploitative. Similar results have been obtained for *Chlorella* in an animal, *Hydra* (Douglas and Smith, 1983). Although further work is required to test how general these findings are for animal-microbial systems, they do indicate that the benefits often assigned to microbial partners in animals (chapter 1, Table 1.1B) should not be assumed, and must be demonstrated empirically.

The finding that hosts may exploit their symbionts in terms of population increase raises important questions about the inclusive fitness consequences of symbiosis for the microbial partners. For many systems, the inclusive fitness needs to take into account the free-living populations. For example, Lowe et al. (2016) had observed *Chlorella* cells in the *Paramecium* culture medium that must have been released from their *Paramecium* host. As discussed in chapter 6 (section 6.2.2), host-associated populations of some microbial taxa may represent a very small proportion of the total population (figure 6.2C). For taxa such as *Chlorella* with large clonal populations of asexually reproducing lineages, the fitness cost of engaging with an exploitative host may, consequently, be trivial. Furthermore, rates of population increase may be an inadequate measure of fitness under natural conditions. For example, the abundance and activities of *Paramecium*, which is promoted by the association with *Chlorella*, may increase the fitness of free-living *Chlorella* populations (for example, by feeding on competitors of *Chlorella*), and the differential between *Chlorella* fitness in symbiosis and in isolation may be strongly contingent on environmental conditions.

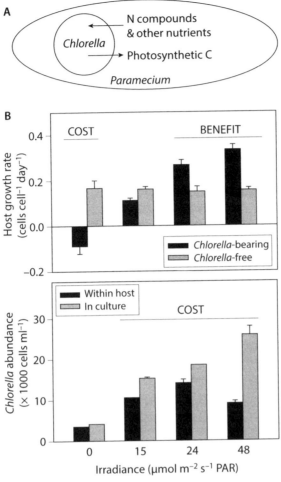

FIG. 7.2. Exploitation of *Chlorella* symbionts by the ciliate host *Paramecium bursaria*. A. Nutrient exchange between the *Paramecium* host and its *Chlorella* symbionts (one *Chlorella* cell of a large *Chlorella* population in each *Paramecium* host is shown). B. Population increase of *Paramecium* (top) and *Chlorella* (bottom) grown either in symbiosis (black) or separately (grey) under continuous darkness (0 irradiance) or 12L:12D light cycle at three irradiance levels. A partner is inferred to derive benefit from the association where its population increase is greater in symbiosis than in isolation, but cost where its population increase is greater in isolation than in symbiosis. (Redrawn from data in Fig. 1 and 2 of Lowe et al. [2016].)

7.2.3. ADDICTION

Some interactions between animals and their resident microorganisms cannot be explained satisfactorily in terms of reciprocity, byproduct mutualism, or exploitation. This can be illustrated by considering germ-free mice, which display many physiological traits that appear to be related in an arbitrary fashion to microbial services (Smith et al., 2007). These mice lack the SCFAs that are released by the gut microbiota of a conventional mouse. A fully predictable consequence of this supplementary source of calories in the conventional mouse is that germ-free mice have reduced fat content, relative to conventional mice, despite equivalent or increased food intake. But the microbial SCFAs have other consequences. For example, they promote tight adhesion of gut epithelial cells, preventing access of some microbial toxins to the internal tissues of the animal, and they stimulate enterochromaffin cells of the gut epithelium to synthesize and release the neurotransmitter serotonin, promoting gut peristalsis (Yano et al., 2015). These responses contribute to the greater resistance of conventional mice than germ-free mice to some microbial pathogens, and to reduced peristalsis and increased transit time of food through the gut of germ-free mice. Why are mice (and presumably other mammals, including humans) dependent on inputs from their microbiota for functions that could be autonomous?

One likely explanation is addiction, i.e., the animal has become dependent on an interaction with the microbiota for a trait, even though the animal (or its ancestors) has the genetic capacity to perform the trait autonomously. Addiction is most likely to evolve in relation to microbial impacts on signaling networks (see chapter 1, figure 1.1). The evolutionary transition to addiction is illustrated in figure 7.3A, together with several routes by which addiction may have evolved, shown in figure 7.3B.

The evolution of addiction by outsourcing function to the microbiota (figure 7.3B-1) has been demonstrated elegantly in a study of interactions among *Drosophila melanogaster*, the bacterium *Wolbachia*, and the pathogenic *Drosophila* C virus (DCV) (Martinez et al., 2016). *Drosophila* has some intrinsic resistance to DCV, mediated largely via one allele of the gene *pastrel*, and *Wolbachia* also confers resistance. Martinez et al. (2016) investigated how the presence of the protective *Wolbachia* may influence selection for the resistant *pastrel* allele. Three replicate *Drosophila* populations, each bearing the same *Wolbachia* genotype or cured of *Wolbachia* with antibiotics, were exposed to DCV in each of 9 consecutive generations (figure 7.4A). As expected for this very harsh selective regime, the evolved flies displayed increased resistance to DCV, relative to flies that not been exposed to DCV over the 9 generations. The evolved *Wolbachia*-free populations, but not the evolved populations bearing *Wolbachia*, displayed an increase in the frequency of the resistant *pastrel* allele (figure 7.4B). The evolved fly populations bearing *Wolbachia* were then cured of *Wolbachia*, and the resultant flies were found to be less resistant to DCV than the evolved *Wolbachia*-free flies, confirming the importance of the resistant *pastrel* allele in antiviral defense. These data show that the antiviral function of the *Wolbachia* impedes the evolution of intrinsic immunity against DCV in *Drosophila*. Essentially, the host has outsourced much of its antiviral

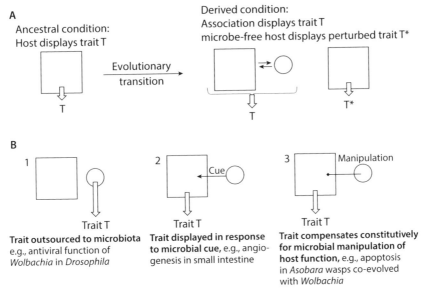

A

Ancestral condition:
Host displays trait T

Evolutionary
transition

Derived condition:
Association displays trait T
microbe-free host displays perturbed trait T*

T

T

T*

B

1

Trait T
Trait outsourced to microbiota
e.g., antiviral function of
Wolbachia in *Drosophila*

2

Cue

Trait T
**Trait displayed in response
to microbial cue**, e.g., angiogenesis in small intestine

3

Manipulation

Trait T
**Trait compensates constitutively
for microbial manipulation of
host function**, e.g., apoptosis
in *Asobara* wasps co-evolved
with *Wolbachia*

FIG. 7.3. Evolutionary addiction and its consequences. A. The evolutionary transition from the ancestral condition with autonomous display of trait (T) by the host (square) to dependence on a relationship with microorganisms (circle) for the trait; in the absence of the microorganisms, the trait is perturbed (T*). B. Mechanisms of addiction, with candidate examples (see text for details).

capability to the *Wolbachia* symbiont. As Martinez et al. (2016) comment, the poor efficacy of immune defenses in hosts bearing a defensive symbiont "may result in the host population becoming dependent on its symbiont to ensure resistance against natural enemies—a form of evolutionary 'addiction' where the symbiont substitutes for host immune defenses."

Other host-microbial interactions may be addictive for the host not because the microbial partner(s) possess the trait of interest but because the presence or activities of the microorganisms provide a reliable cue for the expression of particular traits (figure 7.3B-2). At birth, the small intestine of a mammal bears a relatively sparse capillary network, but a dense network of branching, interconnected capillaries develops in the days following birth, promoting the efficient assimilation of nutrients from the gut lumen and delivery of immune effectors to the gut. This key developmental event is orchestrated by host signaling networks regulating angiogenesis (blood vessel formation). However, the pro-angiogenic signaling cascades in the endothelial and mesenchymal cells of the gut wall require activation by products of microorganisms colonizing the gut (Schirbel et al., 2013), and the microvasculature of the small intestine in germ-free mice does not develop normally and remains sparse throughout life (Stappenbeck et al., 2002). The microbial products are a superbly reliable cue because the developmental switch from nutrition *in*

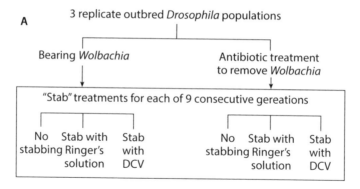

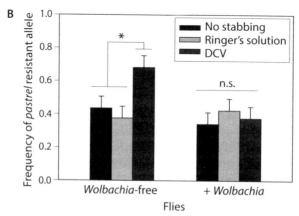

FIG. 7.4. Experimental evolution of resistance to *Drosophila* C virus (DCV) in *Drosophila mela-nogaster*. A. Experimental design. Flies bearing or lacking *Wolbachia* were infected with DCV at each of 9 consecutive generations by dipping a needle into DCV solution and then stabbing the needle into the thorax of the fly. The negative controls comprised flies stabbed with a needle that had been dipped into virus-free Ringer's solution and flies that were not stabbed. B. Frequency of the DCV-resistant allele of the *Drosophila* gene *pastrel* in flies of the 9th generation. (Redrawn from Fig. 1c and 1d of Martinez et al. [2016].)

utero via the placenta to postnatal nutrition via feeding is perfectly coupled to the transition from the microbiologically sterile environment *in utero* to the microbe-rich postnatal environment. (The microbiological status of the developing embryo in mammals, including humans, is discussed in chapter 2, section 2.5.2 and chapter 6, section 6.4.3.) In this way, the timing of post-natal angiogenesis is linked perfectly to the demands for nutrient assimilation across the gut wall. Selection against the dependence of the animal on these pro-angiogenic microbial cues would be minimal because microorganisms are always present, enabling the addiction to evolve. The addiction becomes evident only by experimental elimination of the microbiota.

A further route by which addiction may evolve relates to antagonistic interactions between the animal host and microorganisms: it is in the selective interest of the microbial partner(s) to manipulate signaling networks of the host away from the optimum for the host, and the host compensates for this interaction (figure 7.3B-3). Microbial manipulation of host regulatory networks has been discussed in relation to animal feeding (see chapter 5, section 5.2 and figure 5.2) and may influence many other animal traits. One system offering excellent evidence for addiction arising from compensation for microbial manipulation is the dependence of a parasitic wasp *Asobara tabida* on its *Wolbachia* symbiont. Unlike *Drosophila*, which can readily be cured of its *Wolbachia*, elimination of *Wolbachia* in *Asobara* causes massive apoptosis (programmed cell death) of the insect's ovaries, leaving the wasp reproductively sterile. Several lines of evidence suggest that *Wolbachia* suppresses apoptotic signaling in *Asobara* (Kremer et al., 2009). It appears that, over evolutionary time, the apoptotic pathways in *Asobara* have become increasingly active to compensate for the inhibitory effects of *Wolbachia*, with the consequence that, on removal of *Wolbachia*, the apoptotic machinery goes into overdrive, resulting in the destruction of the ovarian tissues (Pannebakker et al., 2007). In this way, *Asobara* has become dependent on *Wolbachia* symbionts that provide no service. In evolutionary terms, *Asobara* is addicted to *Wolbachia*-mediated manipulation of its apoptotic pathways.

At this juncture, the prevalence of evolutionary addiction in animal-microbial associations is uncertain. Addiction to microbial partners by progressive loss of intrinsic function (figure 7.3B-1) may be widespread in protective symbioses where the microbial partners are consistently present over multiple generations. Addiction through outsourced functions may, similarly, be important for interactions involving microorganisms that mediate social communication, where the animal hosts can become dependent on microorganisms for social communication (chapter 5, section 5.4.3). Indications that addiction via microbial interaction with animal regulatory networks (either figure 7.3B-2 or figure 7.3B-3) may have evolved repeatedly come from the increasing evidence that the presence and composition of microorganisms can alter signaling through many pathways, including insulin, JNK, JAK-STAT, NF-κB, and ERK (Buchon et al., 2009; Kimura et al., 2013; Sansone et al., 2015; Schwabe and Jobin, 2013; Shin et al., 2011), with consequences for many fundamental aspects of animal biology, ranging from cell proliferation, differentiation, and death, through organ-level growth and developmental patterns to the reproductive schedules, biological rhythms, and life history traits of the whole animal.

7.3. Evolutionary Specialization and Its Consequences

7.3.1. PHYLOGENETIC PATTERNS AND THEIR INTERPRETATION

The microbiome of animals tends to be different from the microbial communities in the surrounding environment, and many microbial taxa associated with animals are rare or unknown in the free-living environment (chapter 6, section 6.2). These facts indicate that many animal-associated microorganisms are specialized to the animal habitat, and on an evolutionary trajectory that is distinct from most microorganisms in other environments. Furthermore, the distribution of many microorganisms that are specialized to the animal habitat is restricted to specific taxa, including some instances where the phylogeny of the microbial partner(s) tracks the phylogeny of the host (figure 7.5). These congruent host-microbial phylogenies are of considerable interest from an evolutionary perspective because they may be the result of codiversification of the animal and microbial partners. Furthermore, codiversification provides the opportunity for coevolution, i.e., reciprocal evolutionary change of host and symbiont, which can only proceed if the descendants of one partner are associated with the descendants of the other partner(s). However, as figure 7.5 explains, congruence of phylogenies is necessary but not sufficient evidence

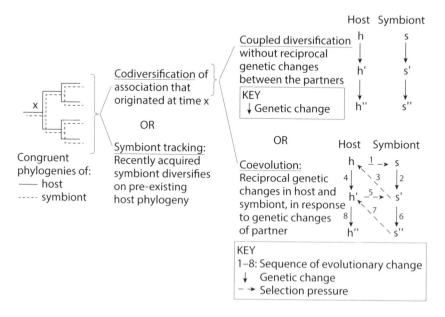

FIG. 7.5. Congruent host-microbial phylogenies and their interpretation. Congruence of phylogenies may be indicative of codiversification or symbiont tracking; and codiversifying partners may have coupled phylogenies (top right) or be coevolving (bottom right).

for codiversification, and codiversification is, similarly, necessary but not sufficient evidence for coevolution.

Of particular interest are the phylogenetic patterns of animals and their microbial partners in open systems, i.e., where individual animals can be colonized by microorganisms from the environment, often throughout their lives. For example, the skin of vertebrates and surface layer of other animals is in continual contact with environmental microorganisms, and the gut of most animals routinely receives microbial inputs with food. With this opportunity for continual gain and loss of microorganisms in open systems, can the phylogeny of any microorganisms map onto the phylogeny of their animal hosts?

Several complementary studies have addressed the relationship between the phylogeny of the animal host and members of the gut microbiota in vertebrates, with particular focus on bacterial taxa that occur in humans. A clear indication that the evolutionary history of some bacteria follows host phylogeny comes from research on the gut bacterium *Lactobacillus reuteri*. Across birds (chicken and turkey) and mammals (mouse, rat, pig, and human), multiple lineages of this bacterial species map to specific hosts and the phylogenetic pattern is firmly indicative of bacterial tracking of a preexisting animal phylogeny (Oh et al., 2010). *L. reuteri* displays functional, as well as taxonomic, diversification, as demonstrated by an experiment in which mice, experimentally deprived of *Lactobacillus*, were challenged with *L. reuteri* from different sources (Frese et al., 2011). With a few exceptions, colonization was limited to the rodent-derived *L. reuteri* (figure 7.6). This pattern can be linked to the localization of *L. reuteri* in the different hosts. In particular, rodent lactobacilli, including *L. reuteri*, form dense biofilms on the epithelium of the forestomach; but in humans *L. reuteri* is generally planktonic and

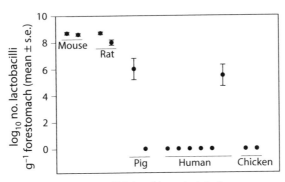

Origin of *L. reuteri* strain

FIG. 7.6. Specialization of strains of the gut bacterium *Lactobacillus reuteri* to their vertebrate host. Mice were experimentally deprived of their gut lactobacilli, but otherwise retained a conventional microbiota, and then administered 10⁶ cells of different *L. reuteri* strains derived from different hosts. The density of lactobacilli in the mouse cecum (not shown) was similar to the density in the forestomach (shown). (Drawn from data in Table 1 of Frese et al. [2011].)

restricted to the colon. Correlated with these ecological differences, many of the rodent *L. reuteri* strains code surface proteins with predicted function in adhesion and biofilm formation, as well as a urease, predicted to confer protection against acidic conditions by its production of ammonia, and IgA-specific protease, protecting against the high IgA exposure in the proximal gut; these genes are generally absent from the human strains of *L. reuteri.*

Congruence of animal and microbial phylogenies is also evident at a much finer phylogenetic scale. A study of the gut microbiota in humans and African great apes (chimpanzees, bonobos, and gorillas) revealed a broad pattern of congruence between the phylogenies of the animal host and both Bifidobacteriaceae and Bacteroidaceae (Moeller et al., 2016), including distinct lineages between sympatric populations of chimpanzee and gorilla. Interestingly, a third family of gut bacteria, the Lachnospiraceae, displayed multiple instances of transfer between host species, and this feature may be associated with the widespread capacity of Lachnospiraceae to form spores that can tolerate conditions in the external environment, while the Bifido-bacteriaceae or Bacteroidaceae do not form resistant spores.

Other studies on the gut microbiota of humans point to effects of host genotype on the prevalence and abundance of certain bacterial taxa. In a study of human twins (Goodrich et al., 2014), there was a significantly greater similarity between monozygotic twins than between dizygotic twins for some taxa, notably the Ruminococcaceae and Lachnospiraceae, but not for others, e.g., Bacteroidaceae. Correlated with this difference, the nodes in the phylogeny of gut bacteria with the highest heritability were within the Ruminococcaceae and Lachnospiraceae, while variation in Bacteroidaceae could be assigned predominantly to environmental factors; and the popula-tions of bacterial groups with higher heritability were more stable over time within individual people than the populations of bacteria with low heritability. Questions arising from this study include the identity of the human genetic factors that influence the prevalence and abundance of some gut bacteria; and the factors that drive different patterns of heritability, including the greater heritability of Lachnospiraceae than Bacteroidaceae in the comparison of different humans (Goodrich et al., 2014) but the reverse pattern in the com-parison across humans and the great apes (Moeller et al., 2016).

An important generality to emerge from these and other studies is that the correspondence between the phylogeny of host and microbial partners varies widely among different members of the same microbial communities, meaning that the host lineage may codiversify with one or more lineages of microorganisms, but not with the microbial community as a whole. This insight facilitates interpretation of an interesting pattern in the composition of microbial communities: for some host lineages, the composition of micro-bial communities (as measured by UniFrac or Bray-Curtis indexes) is more similar between closely related than distantly related hosts (Brucker and Bordenstein, 2013; Franzenburg, Walter, et al., 2013; Ochman et al., 2010). This result, which has been referred to as phylosymbiosis, has been interpreted inappropriately as codiversification of the host and the entire microbial community. A more likely explanation for this pattern is that closely related host taxa support more similar microbial communities because they provide more similar conditions and resources than in distantly related taxa (Douglas and Werren, 2016; Moran and Sloan, 2015). It should also be emphasized that, in many animals, there is no significant phylogenetic signal in the variation in microbiota composition across host taxa, e.g., in *Drosophila* species (Wong et al., 2013) and *Peromyscus* mice (Baxter et al., 2015).

So far, we have focused entirely on open symbioses. Many animal-microbial associations are closed, meaning that the microbial partners are transferred between hosts, usually from parent to offspring, with no opportunity for the microorganisms to escape either to the external environment or to other hosts, nor for the host to acquire alternative microbial partners horizontally. The persistence of these transmission patterns over very many host generations results in codiversification. Codiversification provides the basis for coevolutionary interactions (figure 7.5). Persuasive evidence for coevolution has been obtained for codiversifying associations between insects and microbial partners that have undergone drastic genome reduction, as considered next.

7.3.2. GENOME REDUCTION AND COEVOLUTION

Some insect groups bear vertically transmitted microorganisms that are transferred from the parental host to offspring via the reproductive organs (usually the ovaries of females), with exceptionally long evolutionary histories of codiversification, extending to >100 or even 200 million years. A general feature of these associations is that the bacterial partners have small genomes and much-reduced coding capacity. Genomes are generally <1 Mb (for comparison, the genome of *E. coli* is ca. 4.5 Mb), and some are 0.15–0.3 Mb, in the same range as bacterial-derived organelles (see chapter 2, figure 2.4B). The massive gene loss in these vertically transmitted bacteria can be attributed partly to relaxed selection on genes that are not required in the animal habitat, and also to genomic deterioration, i.e., maladaptive gene loss. Genomic deterioration arises from a key feature of these vertically transmitted systems: relatively small numbers of very closely related bacterial cells are transmitted from parent to offspring. This means that any mutation that arises is likely, by chance, to be present in all the bacterial cells transmitted to an individual offspring. Consequently, selection against the mutation can only operate at the level of whole association, such that hosts with nonfunctional symbionts will be selected against, but mildly deleterious mutations are likely to persist. Over multiple host generations, mutations in the bacterial

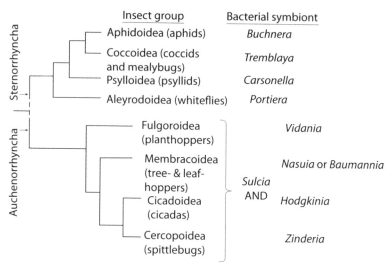

FIG. 7.7. Widely-distributed bacterial symbioses in plant sap–feeding insects (Hemiptera of the suborders Sternorrhyncha and Auchenorrhyncha). Details of the diversity of bacterial symbionts in these insects are available in Douglas (2016).

partner are predicted to accumulate, and this process is accelerated once genes contributing to DNA repair and recombination start to decay (Moran, 1996). Genomic decay is not special to vertically transmitted symbionts, however. As described above, it is caused by population bottlenecking, and it has been described, for example, in the populations of the gut bacterium *Lactobacillus reuteri* in humans, but not other animals (Frese et al. (2011) discuss possible evolutionary scenarios that may explain the unusual genome organization in human-derived *L. reuteri*), as well as some bacterial pathogens, e.g., *Mycobacterium leprae*, and lactic acid bacteria passaged over hundreds of years in traditional yoghurts (Douglas and Klaenhammer, 2010). However, the degree of genomic decay in these latter bacteria is minor compared to many insect bacterial symbionts with genomes <0.5 Mb.

Metabolic coevolution between the vertically transmitted bacteria that codiversify over millions of years with their insect hosts has arisen from the opposing evolutionary processes of symbiont genomic decay and selection for function at the level of the association. Specifically, the host has, repeatedly and in diverse ways, compensated for decay of key metabolic genes in the bacterial symbiont by mediating the reaction. Research on metabolic coevolution has focused largely on animals that feed through the life cycle on plant sap. Across the entire animal kingdom, plant sap (phloem and xylem) are utilized as the sole food source throughout the life cycle only by insects of the order Hemiptera and, within this order, plant sap feeding has evolved multiple times, invariably associated with the possession of microorganisms.

Where tested, as for example in aphids (section 7.2.2), the microbial partners provide the insect host with essential amino acids, which are in short supply in the plant sap. Different groups of these insects have acquired different bacteria, each of which has independently undergone genome reduction and evolved to provide the insect host with essential amino acids. Research on metabolic coevolution has focused on two hemipteran suborders, the Auchenorrhyncha and Sternorrhyncha (figure 7.7).

Let us focus on the associations in sternorrhynchan hemipterans first. The genes for several metabolic reactions that are mediated by many free-living bacteria and also by animals are absent from these symbiotic bacteria. For example, the bacterial symbionts have most of the genes for the synthesis of the branched chain amino acids (BCAs, isoleucine, leucine, and valine) but *Buchnera* in aphids, *Portiera* in whiteflies, and *Tremblaya* in mealybugs have lost the gene mediating the final reaction, branched chain amino acid aminotransferase (BCAT), presumably by genomic decay (figure 7.8A). As in animals, the insect hosts lack the BCA biosynthetic pathway, apart from BCAT. In animals, BCAT generally functions in the reverse direction to mediate the first reaction in BCA degradation, but is reversible and does function in the biosynthetic direction in certain tissues. In aphids, whiteflies, and mealybugs, expression of the insect *BCAT* gene is highly enriched in host cells bearing the symbionts, relative to the whole body, and metabolic experiments conducted on the aphid association confirm that the host cell does indeed mediate the final step in BCA synthesis (Luan et al., 2015; McCutcheon and von Dohlen, 2011; Russell et al., 2013; Sloan et al., 2014). As figure 7.8A illustrates, the BCAs are synthesized by a shared metabolic pathway that has evolved by genomic decay of the bacterial BCAT and recruitment of host BCAT expression to the host cell. The expression of other host genes that have been recruited to shared metabolic pathways with bacterial symbionts are also enriched in the host cells; one example is the proximal reactions in the synthesis of the essential amino acid methionine, linked to the loss of all methionine biosynthesis genes other than *metE*, mediating the final reaction, in *Buchnera* (see section 7.2.1 and figure 7.1C).

The scope of host compensation by differential expression of insect genes is, however, very limited because animals, as a group, have restricted metabolic capabilities. These associations have responded to the selection pressures for function imposed by genomic decay of their symbionts in two ways: fixation of metabolism genes transferred horizontally from bacteria, and acquisition of supplementary bacteria that mediate key metabolic functions (Douglas, 2016). Both of these evolutionary routes are illustrated by the synthesis of the amino acid arginine. (Many animals, including humans, can synthesize arginine at low rates via the urea cycle, but hemipteran insects lack urea cycle genes.) While the *Buchnera* symbiont in aphids has

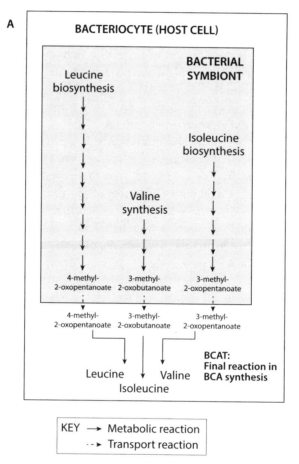

A

BACTERIOCYTE (HOST CELL)

BACTERIAL SYMBIONT

Leucine biosynthesis

Isoleucine biosynthesis

Valine synthesis

4-methyl-
2-oxopentanoate

3-methyl-
2-oxobutanoate

3-methyl-
2-oxopentanoate

4-methyl-
2-oxopentanoate

3-methyl-
2-oxobutanoate

3-methyl-
2-oxopentanoate

**BCAT:
Final reaction in
BCA synthesis**

Leucine   Valine

Isoleucine

KEY → Metabolic reaction
    ⇢ Transport reaction

FIG. 7.8 A. Metabolic coevolution in insect-bacterial symbioses: Loss by genomic decay of the bacterial enzyme branched chain amino acid aminotransferase (BCAT) is compensated by selective expression of the host BCAT enzyme in the host cell. This shared pathway has evolved independently in the aphid-*Buchnera*, mealybug-*Tremblaya* and whitefly-*Portiera* symbioses, but has not been reported in the psyllid-*Carsonella* symbioses.

the genetic capacity for all the reactions, both *Carsonella* in psyllids and *Portiera* in whiteflies lack *argH* mediating the final reaction, and *Tremblaya* in some mealybugs lacks *argF*, the gene for the proximal reaction. As figure 7.8B illustrates, these instances of genomic decay have been compensated for by horizontal gene transfer from bacteria in the psyllid and whitefly symbioses, and by acquisition of a supplementary symbiont in the mealybug symbiosis. The *arg* genes in the whitefly genome are derived from bacteria of the family Enterobacteriaceae, and not the *Portiera* bacterial symbiont (family Halomonadaceae) in these insects (Luan et al., 2015) and the *argH* in the psyllid genome may be derived from the *Carsonella* symbiont (Sloan

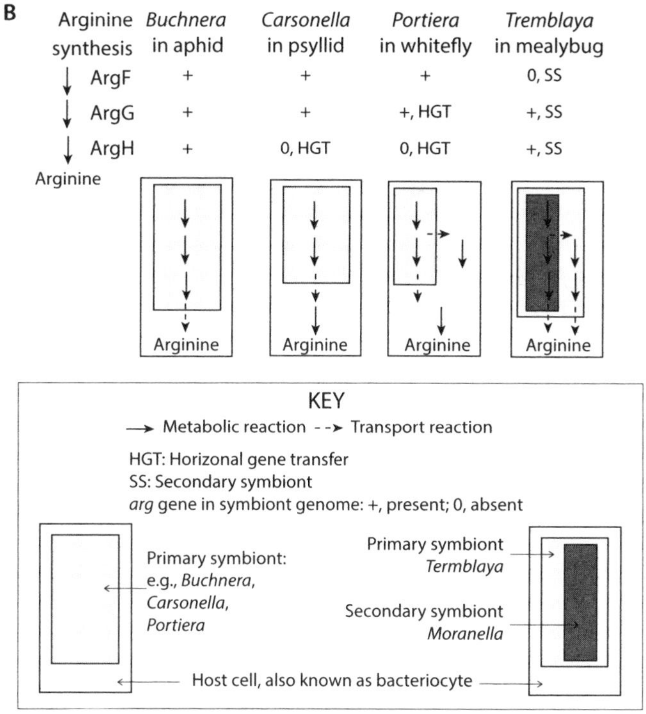

FIG. 7.8B. Metabolic coevolution in insect-bacterial symbioses: Compensation for decay of terminal genes in arginine biosynthetic pathway (*argH* coding arginosuccinate lyase, and *argG* coding arginosuccinate synthase) by genes horizontally transferred from bacteria to the host genome, and by acquisition of a second symbiont in the mealybug. Note that both host and symbiont have the genetic capacity for one reaction (ArgG) in the whitefly, and the two symbionts have the capacity for two reactions (ArgG and ArgH) in the mealybug symbiosis. This metabolic redundancy may facilitate decay of the arg genes in *Portiera* and *Tremblaya*.

et al., 2014); as discussed in section 7.4.2, the putative symbiont source of the psyllid *argH* is most unusual.

As mentioned above, metabolic coevolution between bacteria and animal hosts in sternorrhynchan insects has been driven by unremitting accumulation of deleterious mutations in the bacterial symbionts combined with the unrelenting selection for function imposed by the essential amino acid–deficient diet of plant phloem sap. Similar evolutionary patterns are evident in a second group of hemipteran insects, the auchenorrhynchans (planthoppers, leafhoppers, cicadas, etc.), which have independently evolved the plant sap feeding habit via symbiosis with essential amino acid–producing microorganisms (figure 7.7). In the auchenorrhynchans, the dominant compensatory mechanism for genomic decay has been acquisition of supplementary bacteria that are housed in

C Bacterial symbiosis in cicada (auchenorrhynchan insect)

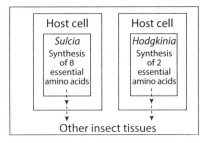

D

his gene	In cicada Tettigades undata	
	Hodgkinia-1	Hodgkinia-2
hisG	+	0
hisI	+	+
hisA	+	+
hisFH	+	+
hisB	+	+
hisC	0	+
hisB	+	+
hisD	+	+

KEY
his gene in Hodgkinia genome: +, present; 0, absent

FIG. 7.8. Metabolic coevolution in insect-bacterial symbioses. C. Partitioning of synthesis of the 10 essential amino acids between two bacterial symbionts localized to separate host cells in auchenorrhynchan hemipteran insects, here illustrated by the cicada symbiosis. D. Partitioning of histidine biosynthesis genes between different lineages of *Hodgkinia* in the cicada *Tettigades undata*. (Data from Fig. 2B of Van Leuven et al. [2014].)

separate host cells and that, in many associations, mediate the synthesis of two essential amino acids, methionine and histidine (figure 7.8C, see also chapter 2, section 2.2.2 and figure 2.2B). A further layer of metabolic coevolution has been uncovered in some species of cicadas, where one bacterial symbiont has diversified into two or more lineages, each of which has lost gene(s) contributing to a key biosynthetic function. Selection for function drives the cotransmission of these complementary lineages from one host generation to the next. Figure 7.8D illustrates the complementary loss of different histidine biosynthesis genes in two co-occurring *Hodgkinia* genotypes in the cicada *Tettigades undata* (Van Leuven et al., 2014). The *Hodgkinia* in another cicada species, *Magicicada tredecim*, has undergone further deterioration into more than twenty genomes of variable size and gene content (Campbell et al., 2015).

In summary, the plant sap feeding hemipteran insects display multiple routes of metabolic coevolution between the animal host and bacterial partners. As more of these associations are being studied, it is becoming evident that coevolutionary changes are ongoing, especially in relation to the horizontal acquisition of different metabolism genes, usually from bacteria other than the symbionts, and acquisition of different supplementary symbionts that contribute metabolic functions decaying in the established symbionts (Douglas, 2016). These associations have evolved over long evolutionary timescales and are highly specialized. The incidence of shared metabolic pathways and metabolic coevolution in other animal-microbial associations with shorter evolutionary histories of codiversification remains to be established.

In this section, I have considered animal-microbial coevolution within one group of animals, the plant sap–feeding hemipteran insects and their vertically transmitted bacteria, for which the evidence for coevolutionary interactions is very strong. Various other animal-microbial associations are described as coevolved in the microbiome literature, often without substantive evidence. In particular, it is extremely unlikely that any animal lineage coevolves with the totality of a complex microbial community, such as the mammalian gut microbiota (see section 7.3.1); and statements in the literature that an individual animal coevolves with its resident microbiota are wrong, contradicting the defining property of coevolution that generations of both partners undergo reciprocal genetic change (figure 7.5). Nevertheless, the incidence and pattern of coevolution in animal-microbial associations is a topic of considerable importance. Coevolutionary interactions may be particularly prevalent in addictive interactions, especially where animal signaling systems may display ever-increasing compensation for escalating manipulation by microbial partners (see figure 7.3B-3). The evolutionary changes in these addictive relationships bear some parallels to the extensively studied coevolutionary arms race between pathogen virulence and host resistance.

7.4. Symbiosis as the Evolutionary Engine of Diversification

7.4.1. DIVERSIFICATION OF MICROBIAL PARTNERS

There is accumulating evidence that association with animals facilitates genetic change in microorganisms, and consequently evolutionary diversification. For bacteria, the principal route is increased rates of horizontal gene transfer (HGT), by which bacteria can acquire novel traits, including the capacity to metabolize novel substrates and detoxify xenobiotics.

A global analysis of the incidence of horizontally transferred genes in bacterial communities from different habitats revealed a 25-fold higher incidence in the genomes of human gut bacteria microorganisms, relative to bacteria from environmental habitats (Smillie et al., 2011). In this study, conducted on >1,000 bacterial genomes from different people, bacteria with the same horizontally acquired genes were significantly more likely to occur in samples taken from the same body site than different body sites, even after correcting for phylogeny, suggesting that co-occurrence in the same animal habitats favor HGT. Other studies have focused on specific genes and functions. For example, some isolates of the human gut bacterium *Bacteroides plebeius* bear genes coding for a porphyranase that are otherwise reported only from marine bacteria. The porphyrin substrate of these enzymes is a complex polysaccharide present in the cell walls of red algae but not terrestrial plants (Hehemann et al., 2012). Intriguingly, *B. plebeius* with the porphyranase genes occurs at high frequency in people in Japan, and it has been suggested that these genes were transferred horizontally to the gut bacteria from marine bacteria, likely associated with red algae eaten as part of the traditional Japanese diet (Hehemann et al., 2010).

Consistent with the high rates of HGT inferred from genome analyses of human gut microorganisms, direct investigations of gene transfer have revealed animal-mediated promotion of bacterial HGT rates. Many of the experimental studies have been conducted on invertebrates, especially soil animals and insects, and focused on the rates of between-bacterial transfer of DNA by conjugation. For example, the presence of earthworms in soil microcosms amended with bacteria bearing a plasmid results in elevated rates of conjugative plasmid transfer to other bacteria, either specific recipient species added to the soil or resident soil bacteria. The recovery of transconjugants only in worm casts (fecal material) and not the bulk soil strongly suggests that passage through the worm gut was required for plasmid transfer (Daane et al., 1997; Thimm et al., 2001). In a similar way, passage through the gut of various insects, including mealworms, houseflies, and fleas, can significantly increase the rate of gene transfer (Aminov, 2011) (see figure 7.9A).

Why should association with animals, especially the animal gut, promote horizontal gene transfer among bacteria? There are several likely contributory factors. Bacterial densities tend to be considerably higher in animal guts than in many other environments (e.g., soil, water), and proximity promotes transfer by cell contact. In addition, the environmental conditions in the gut, including the presence of specific host compounds, may heighten competence for gene transfer. Consistent with the generality that HGT rates among bacteria are elevated under conditions of environmental stress (Aminov, 2011), HGT rates among gut bacteria tend to be increased when the microbiota is perturbed. For example, the hormone norepinephrine, which can attain appreciable levels in the mammalian gut especially in stressed animals, can result in a transient increase in HGT by conjugation in some bacteria (figure 7.9B), and

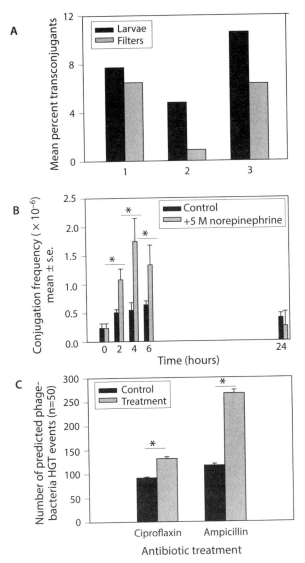

FIG. 7.9. Horizontal gene transfer among bacteria in animal guts. A. Elevated frequency of conjugal transfer of a plasmid from *Salmonella enterica* to *E. coli* in the presence of mealworm larvae *Alphitobius diaperinus* relative to cultures on filters (three replicate experiments 1–3). (Data from Table 4 of Crippen and Poole [2009].) B. Transient increase in frequency of conjugal transfer from *Salmonella to E. coli* in the presence of 5 μM norepinephrine (statistically significant differences are indicated by *) (Redrawn from Fig. 3 of Peterson et al. [2011]). C. Elevated levels of inferred phage-mediated gene transfer among gut bacteria in mice administered the antibiotics ciprofloxacin or ampicillin. (Redrawn from Fig. S8 of Modi et al. [2013].)

gut inflammation caused by a *Salmonella* infection is associated with elevated transconjugative gene transfer between *Salmonella* and *E. coli* (Stecher et al., 2012). Perturbations can also increase levels of phage-mediated HGT among gut bacteria. For example, when the extracellular phage populations in mice treated with the antibiotics ciprofloxacin or ampicillin over 8 weeks were investigated, the representation of phage bearing antibiotic-resistance genes was significantly elevated in fecal samples of the antibiotic-treated mice (Modi et al., 2013), and these genes were confirmed to be functional as demonstrated by greater antibiotic-resistance of bacteria experimentally transfected with phage from these antibiotic-treated mice, relative to phage from control mice. As well as selecting for antibiotic resistance, the antibiotic treatment was associated with increased predicted HGT events between the phage and bacteria (figure 7.9C).

Taken together, these findings indicate that the animal gut is an environment that promotes genetic diversification of bacteria by multiple modes of HGT. The role of the animal gut as a hotspot for bacterial evolution has implications for the diversity of microorganisms that are specialized to the gut habitat, and also for bacteria that only occasionally passage through animal guts. As considered in chapter 6 (section 6.2.2 and figure 6.2C), associations with animals may, in ecological terms, represent sink populations for some bacteria with substantial free-living populations. This apparent disadvantage of associating with animals may, however, be mitigated by the fitness advantage of enhanced gene exchange in the animal environment. These factors need to be taken into account, in relation to pressing biomedical and environmental concerns, including the environmental spread of antibiotic resistance genes and genes from genetically modified organisms. The high connectivity of genes transferred among bacteria in gut communities is predicted to facilitate gene spread among bacteria in different hosts of the same and different species (Aminov, 2011; Smillie et al., 2011; Zurek and Ghosh, 2014), as well as between free-living and animal-associated bacteria (Hehemann et al., 2010). Furthermore, rates of HGT are likely exacerbated by strong selective regimes and perturbed conditions, both of which are generated particularly by antibiotic treatments (Modi et al., 2013).

The significance of the animal gut for evolutionary diversification of microorganisms extends beyond the bacteria to eukaryotic microorganisms, especially ascomycete yeasts. With the increasing appreciation of the importance and diversity of yeasts in many animal-associated habitats, including the human gut and skin (Huffnagle and Noverr, 2013; Oh et al., 2014), these effects may be very widespread but, to date, most of the evidence comes from interactions between the budding yeast *Saccharomyces cerevisiae* and insects. Although *S. cerevisiae* is best known as a model research organism and for its use in baking and brewing, it has a complex natural life style utilizing sugar-rich habitats, especially associated with ripe fruits and sap exudates of trees. Insects that feed on these sugar-rich substrates, e.g., fruit flies and wasps, play an important role in its dispersal and overwintering survival (Chandler et al., 2012; Stefanini et al., 2012).

Association with insects promotes outbreeding in *S. cerevisiae*. The underlying mechanism relates to the life cycle of the yeast (figure 7.10A). When ingested by fruit flies, such as *Drosophila melanogaster*, the vegetative cells are rapidly eliminated, mostly by digestion contributing to the insect nutrition, but the resistant sexual spores remain viable and can be retained in the gut for extended periods (Reuter et al., 2007). However, the capsule enclosing the four products of meiosis of the yeast cell is destroyed in the insect gut (presumably by digestive enzymes, but this has not been established), releasing the individual spores. When these spores germinate, either in the gut or after shedding in feces, they have the opportunity to mate with yeast cells of different genotype, resulting in outbreeding (figure 7.10B). Without the insect-mediated breakdown of the ascus, the germinated spores almost invariably mate with another cell in the same ascus, resulting in selfing.

As mentioned above, *Saccharomyces* yeasts commonly overwinter in the guts of diapausing insects, especially social wasps including hornets and yellow-jackets. Under extended diapause conditions for 4 months or more, many of the yeasts complete the sexual phase of the life cycle, with exceptionally high levels of outbreeding (Stefanini et al., 2016). This outbreeding extends beyond intraspecific crosses within *S. cerevisiae* to include the production of interspecific hybrids. Specifically, when multiple strains of *S. cerevisiae* and its sister species *S. paradoxus* were fed to adult queen wasps of *Polistes dominula* just prior to winter diapause, 25% of the yeasts recovered 4 months later were *S. cerevisiae* x *S. paradoxus* hybrids (figure 7.10C). These results are consistent with abundant genomic evidence for many introgression events between these two species (Hittinger, 2013). Based on the advantageous traits of various interspecific *Saccharomyces* hybrids (e.g., temperature tolerance, growth rates), it is likely that many of the interspecific hybrids produced in the wasp gut would be at a selective advantage, although this was not investigated.

7.4.2. MICROORGANISMS AS A SOURCE OF GENETIC NOVELTY

Associations with microorganisms have brought novel traits to animals, ranging from the synthesis of toxins and nutrients to the degradation of complex dietary constituents and noxious environmental compounds (chapter 1, Table 1.1A). These associations represent one of two routes by which animals have gained access to microbial functions, and the other is HGT. The obvious difference between these two routes is that the microorganism possesses the regulatory and cellular

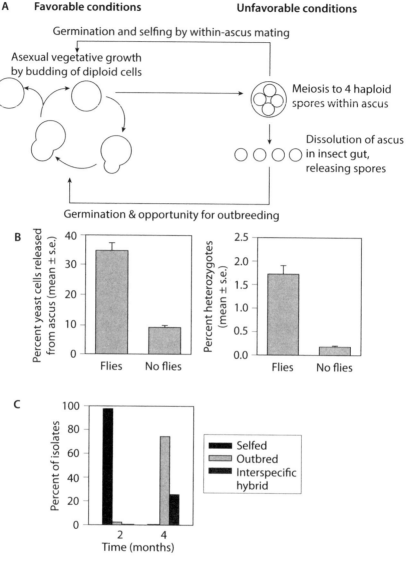

FIG. 7.10. Insect-mediated outbreeding of the yeast *Saccharomyces cerevisiae*. A. The life cycle of *S. cerevisiae*. Under favorable conditions, the yeast cells reproduce vegetatively by mitosis, usually as diploid cells. They respond to unfavorable conditions, such as starvation, by undergoing meiosis to form a tetrad of four haploid spores within a common capsule, known as the ascus; and, on return to favorable conditions, the spores germinate and mate, returning to the diploid condition where they resume vegetative growth by budding. B. Passage of tetrads through the gut of *Drosophila melanogaster* releases many spores from the ascus (left) and significantly increases the heterozygosity of yeast progeny (Fig. S2 and Fig. 1 of Reuter et al. [2007]). C. Incidence of *S. cerevisiae* strains in queen wasps *Polistes dominula* that have overwintered for two and four months (interspecific hybrid refers to *S. cerevisiae* × *S. paradoxus*; pure *S. paradoxus* were not detected in the wasp guts at 4 months). (Redrawn from Fig. 2 A of Stefanini et al. [2016].)

machinery for the expression of the function, while the function of horizontally acquired genes depends on whether and how they are integrated into the regulatory networks controlling expression of the animal genome. Thus, capabilities requiring the coordinated expression of many genes, especially genes with products that assemble into membrane-associated complexes, have particularly been acquired by association with microorganisms. For example, there are multiple evolutionary origins of photosynthesis in animals by symbiosis (Venn et al., 2008) but no animal or any other eukaryotic organism is known to have acquired the independent capacity for photosynthesis by lateral gene transfer. In contrast, there are well-substantiated examples of animal genomes bearing horizontally acquired bacterial genes that mediate functions that are not biochemically complex, e.g., genes that detoxify specific secondary compounds or degrade specific polysaccharides. For example, a β-cyanoalanine synthase of proteobacterial origin in the genomes of some phytophagous mites and lepidopterans (butterflies and moths) mediates detoxification of plant cyanogenic glycosides (Wybouw et al., 2014), and a *Bacillus*-derived mannanase in the coffee borer beetle *Hypothenemus hampei* degrades the galactomannan storage polysaccharide in coffee berries (Acuna et al., 2012). Isolated genes contributing to metabolism have also evolved in coevolved symbioses, most notably in plant sap–feeding insects (see section 7.3.2 and figure 7.8B).

We should not, however, consider intimate associations with microorganisms and HGT as entirely independent sources of evolutionary novelty in animals. Many of the genes of eubacterial origin in the genome of the common ancestor of all modern eukaryotes are derived from the intracellular *Rickettsia*-like symbiont that evolved into the mitochondrion, and the genomes of algae and plants code for many genes derived from the cyanobacterial ancestor of plastids. In other words, close and persistent proximity with an intracellular bacterial symbiont is one of several factors that has facilitated the transfer of many functional genes from bacteria to eukaryotes. These events have had far-reaching evolutionary consequences for eukaryotes, but they occurred in protists.

What is the evidence that animals have gained functionally important traits by HGT from microbial symbionts, in a similar fashion to their protist ancestors? This is an important question to address because symbiotic bacteria are often invoked as the mostly likely source of horizontally acquired genes in animals. HGT to animal genomes may be predicted especially for the plant sap–feeding hemipterans discussed in section 7.3.2 and other insects with microbial symbionts that are vertically transmitted via the reproductive organs. Surprisingly, the scale of HGT from these endosymbionts to the host genome appears to be very limited. Detailed analysis of gene transfer from the γ-proteobacterial symbiont *Buchnera* to the genome of its pea aphid host revealed 12 candidate HGT events from bacteria, of which *Buchnera* was the likely source of just two transfers, both of which are now degraded pseudogenes

(Nikoh et al., 2010). Similar studies of related insects (whiteflies, mealybugs, and psyllids) with different vertically transmitted bacterial symbionts (figure 7.7) have revealed multiple functional genes of bacterial origin in the insect genomes, almost all of which could be assigned unambiguously to bacterial taxa different from the symbiont (Luan et al., 2015); the one exception is the psyllid *argH*, likely derived from the *Carsonella* symbiont (figure 7.8B). Another candidate source of bacterial genes is the bacterium *Wolbachia*, which is widely distributed among insects. Sequences of *Wolbachia* origin have been detected in various insect genomes, including an estimated 30% of a *Wolbachia* genome in the beetle *Callosobruchus chinensis* (Nikoh et al., 2008) and near-complete *Wolbachia* genome in *Drosophila ananassae* (Dunning Hotopp et al., 2007). These *Wolbachia* sequences are, however, generally nonfunctional, with most genes transcriptionally inactive and many pseudogenized.

These data suggest that the evolutionary significance of vertically transmitted intracellular symbionts in animals does not include any substantive role in facilitating HGT from symbiont to animal host. We should, however, consider this conclusion as provisional because the genomes of many insects and other animals are currently being sequenced, and these upcoming data will provide more insight into the scale and significance of symbiont-to-animal gene transfer.

Broadening from this focus on vertically transmitted symbionts to all microorganisms associated with animals, we can conclude that the most likely basis for any impact of symbiosis on animal diversification rates relates to the novel capabilities that the microbial partners bring to the association. Specifically, these innovations and the resultant exploitation of new habitats and lifestyles are predicted to promote evolutionary diversification. In the next section, we turn to consider the evidence that associations with microorganisms are correlated with niche expansion and diversification of animals.

7.4.3. ASSOCIATIONS WITH MICROORGANISMS AS A GENERATOR OF ANIMAL DIVERSIFICATION

The evolutionary diversification and exploitation of new habitats by various animal groups has been linked to the acquisition of certain microorganisms. Consider the bathymodioline mussels, a group of bivalve mollusks that live predominantly in deep-sea habitats of low oxygen content, especially hydrothermal vents and cold seeps, as well as on dead whales and large pieces of wood that fall to the ocean floor (Dubilier et al., 2008). The latter are known as whale falls and wood falls, and represent important sources of organic substrates in the nutrient-poor deep seas (Smith et al., 2015). In many bathymodiolines, the standard mussel feeding habit of filter-feeding is supplemented by single or dual chemosynthetic symbioses with sulfide-oxidizing bacteria and methanotrophic bacteria. Phylogenetic analyses firmly define the ancestral bathymodioline to have lacked this symbiosis and to have utilized organic substrates (wood/whale falls), with multiple evolutionary transitions to the seep and hydrothermal vent habitats (Thubaut et al., 2013). Evolutionary transitions in the reverse direction have not, to date, been reported. Some species utilizing organic substrates have chemosynthetic symbionts, exploiting the sulfide and methane emitted by bacterial decomposition especially of the lipid-rich whale bones; and all seep and vent species have these symbioses (Smith et al., 2015; Thubaut et al., 2013). This pattern gives the likely evolutionary scenario that the whale/wood fall habitat has been the evolutionary cradle of chemosynthetic symbioses, from which the bathymodioline symbioses associated with vents and seeps originated (figure 7.11A). Current data suggest that the taxa in seep and vent habitats are very abundant but not speciose, and the lineages in these habitats have been described as "a kind of evolutionary dead end" (Thubaut et al., 2013).

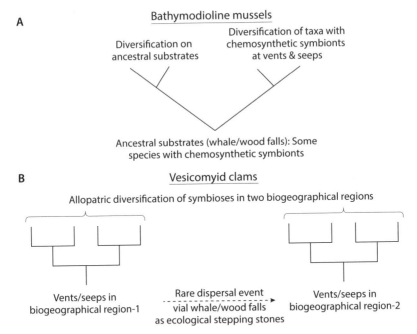

FIG. 7.11. Facilitation of animal diversification by chemosynthetic symbionts. A. Diversification into different habitats. Chemosynthetic symbionts evolved in some bathymodioline mussels utilizing organic substrates (whale/wood falls), followed by the evolutionary transition of some taxa to vents and seeps where these symbioses are required. B. Allopatric diversification in different biogeographic regions. Ancestral chemosynthetic symbioses in one biogeographical region enabled vesicomyid clams to utilize organic substrates during rare dispersal events (perhaps over multiple generations) to different biogeographical regions, followed by allopatric diversification in the different regions. (Redrawn from Fig. 3a and 3b of Smith et al. [2015].)

Chemosynthetic symbioses evolved independently in a different group of bivalve mollusks, the vesicomyid clams. As in the bathymodiolines, symbiosis in vesicomyids has enabled niche expansion and evolutionary diversification of the animal host, but the details are different. The ancestral symbiosis in vesicomyids was most probably associated with seeps and hydrothermal vents, the reverse of the bathymodiolines (Smith et al., 2015), and the diversity of vesicomyid clams dramatically increased 30 million years ago, at the same time as a major diversification of whales into the oceans. Just as some vesicomyid species today utilize both vent and organic substrates, the whale falls may have acted as "ecological stepping-stones," facilitating rare dispersal events among vent/seep habitats (figure 7.11B), thereby enhancing the ecological amplitude and opportunities for evolutionary diversification in these symbioses (Smith et al., 2015).

The inferred relationship between chemosynthetic symbiosis and the niche expansion and evolutionary diversification of their animal hosts shown in figure 7.11 has been reached by a process of reasoned interpretation of species phylogenies. Other systems, however, lend themselves to quantitative analysis, enabling statistical testing of alternative evolutionary scenarios. One system that has attracted particular interest in recent years is a single species, the pea aphid *Acyrthosiphon pisum*, which comprises multiple races that feed from different leguminous plants. Plant affiliation in the pea aphid is significantly correlated with the possession of particular bacteria, known as secondary symbionts. These bacteria are not required by the aphid and they are transmitted maternally with high fidelity, as well as being transmitted horizontally between different aphids. Analysis of the distribution of secondary symbiont taxa in >1,000 pea aphid individuals collected from 13 plant species across different continents (Henry et al., 2013) revealed significant or near-significant associations between plant affiliation and two secondary symbiont taxa, *Hamiltonella defensa* and *Regiella insecticola* (figure 7.12). Furthermore, the genetic relationships among the different aphids indicated that the evolutionary origin of some plant affiliations in the aphid was significantly associated with the horizontal acquisition of a secondary symbiont, suggesting that specific symbionts have facilitated niche expansion of the pea aphid host to utilize novel plant species. For example, multiple independent transitions to utilize *Trifolium* (clover species) in different parts of the world are associated with acquisition of genetically indistinguishable strains of *Regiella* and, similarly, several aphid lineages that have independently adopted *Medicago* (medicks) are colonized by indistinguishable *Hamiltonella* genotypes. Because these bacteria combine vertical and horizontal transmission strategies, they provide a route by which an animal can acquire a complex, functional capability that is heritable. Henry et al. (2013) argue that the relationship between secondary symbionts and pea aphids is,

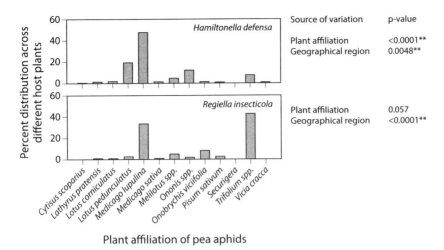

Plant affiliation of pea aphids

FIG. 7.12. Plant affiliation of pea aphids with different bacterial secondary symbionts, *Hamiltonella defensa* and *Regiella insecticola*. Both symbionts displayed significant geographical structuring, and significant or near-significant associations with plant affiliation. (Drawn from data in Table S1 of Henry et al. [2013].)

in this way, analogous to interactions between plasmids and bacterial populations, with both secondary symbionts and plasmids representing a horizontally mobile pool of genes that are advantageous to the aphid or bacteria under certain ecological circumstances.

An important unresolved issue relating to the correlation between pea aphid plant range and secondary symbionts is the underlying mechanism. Because of the very strong statistical support correlating multiple independent origins of plant affiliation with possession of specific bacteria, history (i.e., certain symbionts are coinherited, by chance, with aphid genes mediating plant affiliation) cannot provide a complete explanation. The most robust phenotypic effects of both *Hamiltonella* and *Regiella* relate to protection against natural enemies (parasitic wasps and entomopathogenic fungi, respectively; see chapter 4, section 4.4.2), suggesting that resistance to the natural enemy communities on different plants may be an important evolutionary driver of secondary symbiont-associated plant affiliation. Additionally or alternatively, the symbionts may directly influence the capacity of the aphids to utilize different plant species. However, experiments in which aphids have been infected with different secondary symbionts have yielded inconsistent effects on plant utilization (Hansen and Moran, 2014).

Is the putative role of microorganisms in mediating niche expansion (i.e., utilization of novel plant species) special to pea aphids or widely distributed among animals but poorly recognized? Investigations of microbial determinants of plant range in other insects have yielded mixed results (Hansen and

Moran, 2014). One intensively studied instance of evolutionary divergence linked to shift in plant range is the diverging races of the fruit fly *Rhagoletis pomonella* affiliated with hawthorn and apple. No difference has been identified in the culturable bacteria associated with the two races (Howard et al., 1985), but reanalysis with modern culture-independent methods would be worthwhile. Another system, the cecidomyiid gall midges points to a strong relationship between symbiotic microorganisms and both niche expansion and evolutionary diversification of the animal host (Joy, 2013). As their trivial name suggests, the cecidomyiids induce the plants to form galls, within which the insects develop. Some cecidomyiids are associated with a fungus, *Botryosphaeria dothidea*, which grows on the inner wall of the gall, providing a source of food for the developing midge larvae. Phylogenetic analysis of these insects has revealed significantly higher rates of diversification of lineages with the fungal partner than of those without, and this diversification is accompanied by niche expansion to utilize 7-fold more plant species than the fungus-independent taxa.

The role of symbiotic microorganisms in promoting diversification of their animal hosts is predicated on the speciation of the animals containing these microbial symbionts, often at accelerated rates. These considerations have led to the proposal that the microbial symbionts may, repeatedly, play a direct role in promoting barriers to gene flow and speciation of their animal hosts. This claim is not parsimonious: the traits conferred by the microbial partner can enable the animal host to adopt new habits, with subsequent speciation events mediated by the same suite of processes as for symbiosis-independent speciation. However, as pointed out in relation to the evolution of symbioses generally (O'Malley, 2015), evolutionary processes are not necessarily parsimonious, and various studies have revealed candidate examples of symbiosis-mediated speciation.

7.4.4. SYMBIOSIS-MEDIATED SPECIATION OF ANIMALS?

Until recently, consideration of symbiosis-mediated speciation was restricted to generalities, including claims that the underlying processes are, somehow, distinct from the microevolutionary changes driven by mutation, recombination, selection, and genetic drift (see review of O'Malley, 2015). This area has, however, been reinvigorated by recent studies mostly conducted on insects that offer potential mechanisms of symbiont-mediated interruption to gene flow. The putative mechanisms of reproductive isolation are diverse, and include both prezygotic (i.e., before mating) and postzygotic processes.

Prezygotic isolation is driven by mating behavior, specifically behavioral decisions based on the recognition of species, sex, genetic relatedness, and often group identity. Microbial symbionts can influence these behavioral decisions by their chemical communication among conspecific animals. As considered in chapter 6 (section 6.4), there is a growing body of evidence that some animals utilize volatiles of microbial origin as aggregation pheromones and to identify fellow group members. In principle, if the mating decisions of the animals were dictated by microbial products, and the microbiota is inherited faithfully within the diverging animal lineages, the resultant interruption to gene flow could, in due course, lead to speciation. Supportive empirical data come from experiments conducted on *Drosophila melanogaster* fruit flies, revealing preferred mating with flies of the same gut microbiota complement (Sharon et al., 2010). This experiment used *Drosophila* that had been reared on different diets, yielding flies with either a diverse gut microbiota (corn-molasses-yeast diet) or *Lactobacillus*-dominated microbiota (SY, starch-yeast diet) (figure 7.13A). When offered the choice between mates with the same or different gut microbiota, the flies significantly preferred to mate with flies bearing the same gut microbiota (figure 7.13B). This effect was abolished by antibiotic treatment, and was rescued by colonizing the antibiotic-treated flies with bacteria from the two food types. Furthermore, assortative mating was also recovered by colonization with a single bacterium, *Lactobacillus plantarum*, derived from the SY-reared flies. It is not understood fully how the *Lactobacillus* mediates mating preference, nor whether other bacteria contribute to the assortative mating in conventional flies. However, there are indications that the composition of the gut microbiota may influence the composition of cuticular hydrocarbons, which are well-known to function as mating recognition cues in *Drosophila* and many other insects.

The results of Sharon et al. (2010) are important for providing proof-of-principle that animal-associated microorganisms can promote assortative mating. Their significance for speciation depends on the constancy of the gut microbiota across multiple host generations. This may require rather special circumstances because laboratory cultures of *D. melanogaster* display considerable variability in microbiota composition over time (Wong et al., 2013), much of which appears to be stochastic. Similar variability is likely in field populations. Thus, the mechanisms identified by Sharon et al. (2010) may prove not to be significant in the diversification of *D. melanogaster* and related species, but could be of defining importance in other animals with different microbial dynamics.

We will return to microbial determinants of prezygotic isolation in *Drosophila* at the end of this section but, first, let us consider the role of microorganisms in postzygotic reproductive isolation of animals caused by hybrid unfitness and evident as reduced fertility or poor viability. Although many instances of hybrid unfitness can be explained entirely in terms of breakdown of coadapted gene complexes specific to each parental species, high mortality

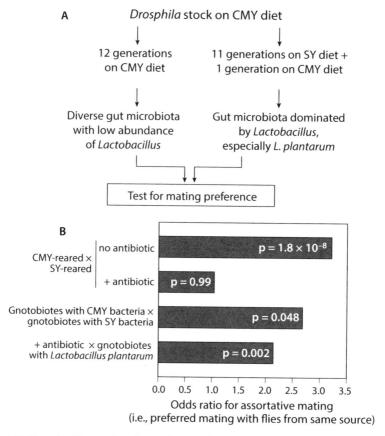

FIG. 7.13. Gut microbiota and mating preference of *Drosophila melanogaster*. A. Experimental design: stocks raised on alternative diets, CMY (corn-molasses-yeast) and SY (starch-yeast) yield flies with different gut microbiota. B. Statistically-significant assortative mating (p<0.05) between conventional flies of different dietary history and gnotobiotic flies with different microbial complements. The gnotobiotic flies comprise *Drosophila* cleared of the microbiota with antibiotics and then colonized with bacteria isolated from CMY diet or SY diet, or with *Lactobacillus plantarum* isolated from flies reared on SY. (Drawn from data in corrected Table 1 of Sharon et al. [2010].)

of some hybrids has been linked to perturbations of the microbiota. Involvement of the microbiota is particularly clearly demonstrated by research on jewel wasps of the genus *Nasonia*. Interspecific crosses between two species *N. vitripennis* and *N. giraulti* yield viable and fecund daughters (generation F_1) but high larval mortality of grandsons (generation F_2). Genetic studies have attributed this to incompatibility between maternally inherited factors and the hybrid nuclear genome (Breeuwer and Werren, 1995). However, the hybrid lethality is abolished by rearing the insects with antibiotic, and is reinstated by adding back two bacteria previously isolated from the insects,

Providencia sp. and *Proteus mirabilis* (figure 7.14) (Brucker and Bordenstein, 2013). Although the cause of death has not been established definitively, the hybrid males appeared to be unable to control the populations of the *Proteus* bacterium, which is a minor member of the gut microbiota in both parental species but dominates the microbiota of hybrids (Brucker and Bordenstein, 2013). In other words, the hybrid suffers dysbiosis, which may be caused by generalized unfitness or possibly specific lesions in the immune system of the hybrids (Chandler and Turelli, 2014).

How general is dysbiosis of the gut microbiota as a factor contributing to hybrid unfitness? At present, we do not know because, apart from a few early studies focusing on specific culturable taxa, reviewed by Shropshire and Bordenstein (2016), the microbiota of hybrids and their parental species have rarely been compared systematically.

A complementary route for microbial mediation of hybrid unfitness is cytoplasmic incompatibility (CI), which is restricted to a few vertically transmitted bacteria, most notably *Wolbachia*. In CI, crosses between males infected with a *Wolbachia* and either uninfected females or females

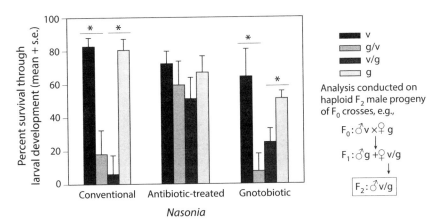

FIG. 7.14. Microbe-dependence of hybrid lethality in jewel wasps, *Nasonia*. *Nasonia* is haplodiploid (males are haploid and develop from unfertilized eggs, while females are diploid and develop from fertilized eggs). Hybrid lethality (left) was scored as mortality of haploid males in the F_2 generation following within-species and between-species crosses between *N. vitripennis* (v) and *N. giraulti* (g), (right). It has been argued that lethality is restricted to males because hybrid lethality genes tend to be recessive, with phenotypic effects evident only in haploid male condition. Adult *Nasonia* females lay eggs in pupae of other insects, including *Sarcophaga bullata*, where they develop through the larval stage to adulthood. The microbiological treatments comprised: conventional (*Nasonia* with an unmanipulated microbiota reared in natural host insect *S. bullata*), antibiotic-treated (*Nasonia* reared on sterile food containing antibiotics, to eliminate the microbiota), and gnotobiotic (following antibiotic-treatment, the *Nasonia* were reared on sterile food inoculated with *Providencia* and *Proteus* bacteria (1:1 suspension) isolated previously from *Nasonia*). (Redrawn from Fig. 1C of Brucker and Bordenstein [2013].)

bearing a different *Wolbachia* are inviable, with death at an early stage of embryogenesis (figure 7.15A). CI is caused by a delayed chromosomal condensation of the paternal chromosomes in the first mitotic division of the zygote, linked to a poorly understood modification to the chromatin during spermatogenesis in *Wolbachia*-bearing males. CI promotes the spread of

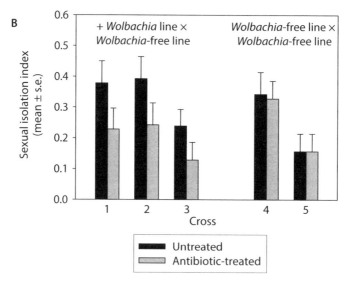

FIG. 7.15. Interruption to gene flow by *Wolbachia*-induced cytoplasmic incompatibility (CI). A. CI in unidirectional (left) and bidirectional (right) crosses (*Wolbachia*-infected individuals indicated by w-subscripts). B. Assortative mating, scored as sexual isolation index (difference between the number of matings between partners from the same population and different populations, as proportion of total number of matings) between *Drosophila melanogaster* lines. The antibiotic treatment was administered 6 generations before assay and was confirmed to eliminate *Wolbachia*. Each cross (1–5) comprises flies from different pairs of *Drosophila* lines; the two *Wolbachia*-free crosses (4, 5) provide a control for nonspecific effects of the antibiotic on mating preference. (Redrawn from Fig. 2 of Koukou et al. [2006].)

Wolbachia-bearing females, which is why *Wolbachia* is often described as a reproductive parasite.

Bidirectional CI is of particular interest because it can, in principle, lead to "rapid speciation": crosses between individuals with different *Wolbachia* produce no viable offspring (figure 7.15A), thereby interrupting gene flow between the parental lines. CI can, in turn, lead to selection for assortative mating, whereby females discriminate against *Wolbachia*-infected males, with the implication that *Wolbachia*-induced postzygotic incompatibility may be obscured by more rapid evolution of prezygotic isolation. Data from long-term population cage experiments on *Drosophila melanogaster* support this view (Koukou et al., 2006). Over a period of just five years, flies from the separated populations evolved significant preference to mate with partners from the same population, relative to different populations. In mating preference trials between flies from populations in which *Wolbachia* was either fixed or absent, the preference index was reduced by 50% if the *Wolbachia* had been eliminated by antibiotic treatment, but antibiotic-treatment did not affect the mating preference of *Wolbachia*-free crosses (figure 7.15B).

The conclusion from this study is that the promotion of postzygotic isolation by *Wolbachia* can select over relatively short timescales for assortative mating and prezygotic isolation. Data for natural populations of other *Drosophila* species, including different populations of *D. paulistorum* and between *D. recens* and *D. subquinaria*, as well as other *Wolbachia*-infected arthropods suggest that this may occur in natural insect populations (Brucker and Bordenstein, 2012; Shropshire and Bordenstein, 2016). Indeed, it can be argued that interruption to gene flow caused by *Wolbachia*-mediated incompatibilities may evolve more rapidly, and be more widespread than nuclear incompatibilities (and associated dysbiosis) because the fitness cost of *Wolbachia* is dominant and evident within a single generation (figure 7.15A), whereas other incompatibilities between hybrids tend to be recessive and are expressed in the F_2 generation (e.g., see figure 7.14).

7.5. Summary

The association between microorganisms and animals has immense evolutionary consequences at scales ranging from the fitness of participating organisms to the rates and patterns of evolutionary diversification of both animals and their microbial partners.

Many associations are believed to be founded on reciprocity, i.e., the reciprocal exchange of services that confer greater benefit to the recipient than cost to the donor (section 7.2.1). Multiple ways in which animals benefit from gaining access to metabolic capabilities of microorganisms have been established, but the advantage of associating with animals to the microbial partners is generally described nonquantitatively in terms of the animal as a habitat that confers access to nutrients or enemy-free space. The possibility that some animal hosts may exploit their microbial partners (section 7.2.2) is raised by a study demonstrating that *Chlorella* algae incur net cost from their association with the ciliate *Paramecium bursaria*. The likely significance of the reverse, exploitation of animals by their microbial partners, is indicated by patterns of host-symbiont metabolic interactions that constrain both free-loading (failure to provide a service) and overconsumption by the symbiont.

Many microbe-dependent traits of animals cannot readily be explained in terms of microbial services, but relate to microbial impacts on the signaling networks that regulate animal functions required for sustained animal health. Some of these interactions are likely to be addictive, i.e., the animal is dependent on the microbial partner(s) for a trait even though it has the genetic capacity to perform the trait autonomously (section 7.2.3). Three types of addiction can be identified: the outsourcing of a trait to the microbiota, dependence on a microbial cue for appropriate expression of the trait, and constitutive compensation for microbial manipulation of a trait.

Many animals and their microbial partners display congruent phylogenies, such that microorganisms associated with closely related hosts are more closely related than microorganisms in phylogenetically distant hosts (section 7.3). Congruent phylogenies can be generated by diversification of the microbial partner onto a preexisting host phylogeny (as has occurred in *Lactobacillus reuteri* associated with vertebrate guts) or by codiversification. Coevolution, i.e., the reciprocal genetic changes in host and microbial partners, has been demonstrated between plant sap–feeding insects and vertically transmitted bacterial symbionts, but its wider incidence in microbial associations with healthy animals is uncertain.

As well as shaping the patterns of diversity, associations between microorganisms and animals can influence the rate of diversification of all partners (section 7.4). There is increasing evidence that residence in animal guts can promote microbial diversification via horizontal gene transfer of bacteria and out-crossing in yeasts, and that the rates of diversification in animal clades can be elevated in lineages bearing microbial symbionts relative to symbiont-free clades (e.g., in gall midges bearing and lacking fungal partners). The processes underlying diversification rates may be diverse, and could include direct effects of microbial partners and their products on mating preference and host viability. However, further research is required to assess the prevalence and mechanism of microbe-mediated speciation in animals.

8

The Animal Reimagined

8.1. Introduction

This book provides a survey of the state-of-the-art for microbiome science with a focus on six broad areas:

The evolution of animal microbiology from a historical perspective as an explanation for how it all began (chapter 2), and evolution from a mechanistic perspective, offering explanations of ongoing processes that are shaping the structure of animal microbiomes (chapter 7);

the impacts of the microbiome on human health (chapter 3), reflecting the central role of biomedical science in the development of the discipline of microbiome science;

interactions between the microbiome and two physiological systems of animals: immunity (chapter 4), supported by a very substantial body of research, and behavior (chapter 5), which has attracted much discussion but rather less empirical research; and

the key ecological concepts starting to be applied to microbiome science (chapter 6), in recognition of the power that the discipline of ecology can bring to understand complex microbiome-animal interactions.

In this chapter, I consider how an appreciation of the microbiology of healthy animals is changing our understanding of animals and I identify some future priorities for microbiome science. Scientists are no better than astrologers at crystal ball–gazing, and I write this chapter with some trepidation. After all, the otherwise superlative book of Paul Buchner on endosymbiosis in animals

(Buchner, 1965) misstepped just once, with the recommendation that research on the symbiotic origin of organelles would be a "wrong path" to take.

The three topics selected for this chapter are the pervasive impact of the microbiome on animal biology (section 8.2), the priority to include the microbiome in explanations of the phenotype of animals (section 8.3) and, finally, the role that targeted research on the microbiome can play to enhance our capacity to mitigate and manage some unintended negative consequences of human activities in the Anthropocene (section 8.4).

8.2. The Scope of the Animal

8.2.1. A PHYLOGENETIC PERSPECTIVE

All animals are associated with microorganisms. Our ancestors were multiorganismal before they were multicellular (chapter 2), and animals, including humans, have diversified and live their daily lives in the context of persistent interactions with microorganisms that are adapted to varying extents to the animal habitat (chapter 3 and chapter 7).

Overlaying this generality are several broad patterns in the phylogenetic distribution of different types of associations. A particularly striking phylogenetic pattern relates to photosynthetic symbionts, which are mostly restricted to basal animal groups, especially the sponges and cnidarians (corals and relatives) and flatworms, while with a few exceptions, e.g., some sea slugs and the giant clams, the photosynthetic lifestyle is largely unknown in other animals. Another broad functional pattern relates to the herbivorous lifestyle. For vertebrates and especially endotherms (birds and mammals), feeding on living plant material is very strongly associated with the possession of complex communities of cellulolytic gut microorganisms in a gut fermentation chamber, but equivalent associations in invertebrate animals are restricted to taxa utilizing recalcitrant plant material, e.g., wood-feeding termites.

The incidence of intracellular microorganisms provides a further distinctive phylogenetic pattern. In animals with relatively few cell types, intracellular symbionts occur in cells with other functions (e.g., *Chlorella* algae in the digestive cells of *Hydra*) but in morphologically complex animals, they tend to be restricted to specialized animal cells, whose sole function appears to be to house and maintain the microbial cells. This specialization is particularly evident in insects, where the associations have evolved independently multiple times and persisted with obligate vertical transmission for many millions of years (Buchner, 1965; Douglas, 2015). In sharp contrast, the vertebrates almost entirely lack intracellular microorganisms other than overt pathogens. Just one known intracellular endosymbiont is known, the alga *Oophila* in embryos of the spotted salamander *Ambystoma maculatum* (Kerney et al.,

2011). The lack of intracellular symbionts in vertebrates may have constrained the levels of dietary or other habitat specializations in this group, especially by comparison to the ancient coevolved associations with intracellular symbionts that have enabled various insect groups to live on extremely unbalanced diets (chapter 7, section 7.3). Most animal groups have not been studied as intensively as the insects and vertebrates, and the overall incidence in animals of ancient intracellular microorganisms is not well known.

Various explanations for the observed patterns have been put forward: that the simple body plans of cnidarians and flatworms are well suited to the morphological modifications facilitating light capture for photosynthesis (Venn et al., 2008); that the adaptive immune system enables the management of taxonomically complex microbial communities in the vertebrate gut (McFall-Ngai, 2007), but constrains the evolution of intracellular symbioses (Douglas, 2010); and that the dependence of vertebrate herbivores on a cellulolytic gut microbiota may be related to the loss of the genetic capacity to degrade cellulose in the lineage giving rise to vertebrates, compounded by the high energy demand for endothermy in birds and mammals (Karasov and Douglas, 2013). In all of these various explanations, differences in the relationship between the animals and their microbiota are treated as consequences of microbe-independent evolutionary changes, especially increases in morphological or functional complexity. As the discipline of microbiome science matures, including toward a more comparative science (i.e., phylogenetically informed comparisons across animal taxa), explanations for phylogenetic patterns are likely to become more sophisticated and open to empirical testing. The studies of the relationship between associations with symbionts and the rates and pattern of host diversification described in chapter 7 (section 7.4.3) illustrate the potential of these approaches.

8.2.2. PHYSIOLOGICAL SYSTEMS

With the ample evidence that the abundance and composition of resident microorganisms in animals varies between different regions of the animal body and is tightly regulated, principally by the immune system (chapter 3, section 3.2; chapter 4, section 4.2), we can reasonably predict major effects of the microbiota on the cells, tissues, and organs with which they are associated. In humans, the principal foci for microbiome interactions are the mouth and gut, the skin and the respiratory tract, as well as the vagina of women. The composition and activities of the microbiota in these locations are correlated with healthy or diseased states of these organs, including atopic dermatitis and psoriasis of the skin (Zeeuwen et al., 2013), gum disease and dental caries (Wade, 2013), inflammatory bowel disease and some cancers of the GI tract (Sears and Garrett, 2014), cystic fibrosis and COPD of the lung (Dickson et al., 2016), and susceptibility to HIV and other sexually transmitted diseases in the vagina (Cohen, 2016).

It has become clear, however, that the effects of the microbiome extend beyond the organs with which they are associated. The multiple physiological and biochemical differences between a germ-free mouse and conventional mouse illustrate how the effects of the microbiome are pervasive to all physiological systems (Smith et al., 2007). Many of the effects at a distance can be attributed to the distribution of microbial products, including microbial metabolites and cell wall fragments, to different regions of the body, especially via the circulatory system. For example, bacterial short chain fatty acids and peptidoglycan fragments that escape from the gut to the blood system have been implicated in regulation of the mammalian circadian rhythm of metabolism and sleep, respectively (chapter 5, section 5.3.4; chapter 6, section 6.4.4). Other effects are mediated through local host responses to microbial effectors. Thus, some of the likely effects of the gut microbiota on the brain and behavior can be attributed to interactions between microbial products and immune cells and neurons within the gut wall, with animal-mediated feed-forward effects to the brain (chapter 4, section 4.3.2 and chapter 5, sections 5.2 and 5.3).

These considerations lead to a general expectation that animal physiological systems are subject to the influence of the microbiota. The consequences are substantial. There are many potential opportunities for novel microbiome-informed therapies for a diversity of diseases (Rajpal and Brown, 2013), and microorganisms offer both a vehicle and a target for novel strategies to control pests and disease vectors (Douglas, 2015). Furthermore, an awareness of the microbiome can contribute to effective management of transmissible diseases, as illustrated by the finding that transmission of *Plasmodium*, the agent of malaria, by *Anopheles* mosquitoes is facilitated by feeding on people taking antibiotics that clear the mosquito gut of *Plasmodium*-suppressive microbes (Gendrin et al., 2015). A further and far from trivial consequence is the extensive rewriting of the next edition of every physiology textbook and undergraduate lecture course to accommodate the pervasive role of the microbiome in animal biology.

The therapeutic impact of microbial therapies to modulate the function of physiological systems for improved health remains uncertain, in terms of both the magnitude of the effect and range of disease conditions that may be responsive (chapter 3, section 3.5.3). Two factors may limit responsiveness. First, some microbiomes may be exceptionally robust, either taxonomically or functionally, to perturbation. Currently, rather little is known about the incidence of alternative stable states of the microbiome in humans and other animals, and their significance for health (chapter 6, section 6.4.2), and there is much work to be done in this regard. Second, some aspects of the animal biology may be functionally insulated from the microbiota. For example, the serotonin titer in the brain, which plays important roles in feeding behavior and mood, is metabolically isolated from the effects of the gut microbiota on serotonin production in the gut (chapter 5, section 5.2.3). Another aspect of the biology of the healthy animal that has traditionally been treated as isolated from microbiological influence has been embryogenesis and development, but, as considered next, the relationship between the microbiome and development is a further likely area of very productive future research.

8.2.3. DEVELOPMENTAL SYSTEMS

In most animals, embryogenesis appears to be perfectly isolated from the microbiome, in the sense that embryogenesis proceeds normally in the absence of microorganisms. The surface coverings of the deposited eggs of many animal species bear microorganisms, often of maternal origin. These microorganisms may function to produce toxic compounds that protect the egg against predation (Florez et al., 2015; Gil-Turnes et al., 1989) and provide an initial inoculum for the hatched offspring, but their removal does not affect embryo development (chapter 6, section 6.2.1). Where studied, embryogenesis in viviparous species is, similarly, independent of microorganisms. Although it is debated whether embryo development in mammals, including humans, proceeds under completely sterile conditions in utero (chapter 2, section 2.5.2), the abundance and density of microorganisms is trivial compared to other colonized locations in the healthy human (figure 3.1). The only animals with embryogenesis in close proximity to a substantial microbiota are the species with vertically transmitted microorganisms transferred into the cytoplasm of the developing oocyte. These systems have evolved in various insects, and experimental studies have generally revealed no requirement for normal embryogenesis. One apparent exception relates to an early study on the leafhopper *Euscelis incisus* where treatment with high antibiotic concentrations caused abnormal development of the abdomen. The interpretation that the microorganisms may define the anterior-posterior axis of this insect (still occasionally referred to in review articles) has since been refuted by the demonstration that the developmental abnormalities were caused by toxicity of the antibiotic and that elimination of the microorganisms by mechanical methods caused no developmental defects or abnormalities during embryogenesis (Douglas, 1988).

The evidence against a general role of the microbiome in embryogenesis should not be confused with the abundance of evidence that resident microorganisms interact with the developmental programs of postnatal animals at multiple levels (McFall-Ngai, 2002). Examples of great significance for mammalian health have been discussed in previous chapters of the book, including the angiogenesis of the mammalian gut and maturation of lymphoid cells of the immune system (chapter 4, section 4.3.3 and chapter 7, section 7.2.3), as well as evidence for microbial contributions to the regulation of gut epithelial differentiation and proliferation in both mammals and insects (Bates et al., 2006; Broderick et al., 2014; Yu et al., 2015).

There are also starting lines of evidence that microbial products of maternal origin may modulate specific aspects of the developmental program of offspring. In particular, bacterial peptidoglycan administered to the circulatory system of pregnant mice can traverse the placenta to the developing brain, causing neuroproliferation and cognitive dysfunction of the offspring (Humann et al., 2016); and microbial products, including peptidoglycan and LPS, can be transferred to the eggs of the honeybee in association with the vitellogenin yolk proteins, providing immune-priming molecules that influence the immunological function of the off-spring (Salmela et al., 2015). These examples relate to the pathological effects of microbial products and transgenerational protective responses against pathogens, respectively, but they provide the proof-of-principle for possible maternal effects mediated via metabolic or other products of the microbiome, without direct interaction of microbial cells of maternal origin with the developing offspring.

These data suggest that the priority for future research on the inter-actions between the microbiota and animal developmental systems should include (1) products of the maternal microbiota that may influence the pattern of embryogenesis in the offspring, and (2) early postnatal devel-opment, when newly acquired microorganisms provide cues for the mat-uration of various organs and physiological systems. Many of these effects are likely to be subtle, but they could also include major developmental switches, for example in animal species with different morphs, e.g., the presence of certain bacteria in aphids has been associated with wing polyphenism (Leonardo and Mondor, 2006) although the underlying mechanism is not known. Involvement of microorganisms or their products in key developmental events also raises questions about the implications of the likely reduced diversity of the microbiome associated with modern lifestyles. A microbiome of low taxonomic or functional diversity may provide inappropriate or insufficient stimulation at key developmental stages, with health consequences in later life. These considerations rep-resent an extension of the Barker hypothesis that some chronic adult diseases have their origins during fetal development (Almond and Currie, 2011; Barker, 1992).

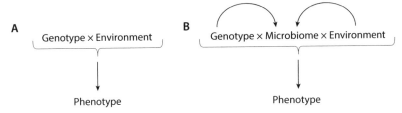

FIG. 8.1. Updating G × E. A. In the premicrobiome era, animal phenotype was interpreted to be determined by the interaction between genotype and environment. B. Animal phenotype is shaped by the three-way interaction between genotype, microbiome, and environment, with the microbiome influenced by both animal genotype and environment.

8.3. The Determinants of Animal Phenotype

8.3.1. UPDATING G X E

The phenotype of an animal is shaped by the interaction between genotype and environment, commonly abbreviated to G X E (figure 8.1A). G X E interactions have been the mainstay of population genetics, life history theory, and some aspects of biomedical science for decades. The obvious difficulty is that G X E omits the microbiome, which has major effects on the animal phenotype.

The complexity of the determinants of animal phenotype are complicated further by the evidence that the composition and activities of the microbiome are strongly influenced by both host genotype and the environment (figure 8.1B). Multiple studies have demonstrated environmental effects on the microbiome, most notably diet and temperature (see chapter 6, sections 6.4.1 and 6.4.2; and reviews of Candela et al. [2012] and Wernegreen [2012]). The effect of host genotype on the microbiome has, to date, attracted far less attention, but genetic factors are undoubtedly important, as demonstrated, for example, by research on the gut microbiota of *Drosophila* (Chaston et al., 2016). Analysis of 79 *Drosophila* lines revealed considerable among-genotype variation in colonization by a gut bacterium *Acetobacter pomorum*. This variation was correlated by genome-wide association study (GWAS) to several animal genes, particularly genes expressed in the CNS; and these results may reflect controls over the microbiota via the gut-brain axis, with significant phenotypic consequences for the flies. Mutants of two genes tested significantly affected fly phenotype, but only when the gut microbiota included *A. pomorum*. This study illustrates how inclusion of the microbiome and its interactions with both the host genotype and environmental variables can enrich our understanding of the determinants of the phenotype of animals.

8.3.2. HOW ADAPTIVE ARE PHENOTYPES MEDIATED BY MICROBIAL INTERACTIONS WITH G X E?

In many systems, the effects of environmental factors on the microbiota are adaptive for the animal host. When desert woodrats *Neotoma lepida* were fed on a toxin-laden creosote bush *Larrea tridentata*, their gut microbiome changed to a composition with greater capacity to metabolize these toxins (Kohl et al., 2014); germ-free mice administered the gut microbiota from cold-adapted mice had a greater capacity to tolerate cold conditions than mice colonized with the microbiota from warm-adapted mice (Chevalier et al., 2015); and corals that modify their algal symbionts in response to high temperature are more tolerant of subsequent thermal stress than corals with unaltered algal symbionts (Bay et al., 2016).

These examples illustrate that the mechanisms regulating the composition and activities of the microbiome can be integrated with other adaptive responses of animals to environmental circumstance, reinforcing the evidence (section 8.1.2) that the microbiome is strongly incorporated into the physiological systems of animals. Although the processes by which the animal partner responds to environment-correlated changes in the microbiota are increasingly being studied, the mechanisms by which environmental factors influence the microbiota, either directly or interactively with host responses to the environmental factors, are less well understood.

In some instances, however, the microbiota X environment interactions may not be in the selective interest of the animal host. Let us consider the likely compounding effects of inappropriate diet and dysbiotic microbiota on human nutritional physiology, causing metabolic disease (chapter 5, section 5.2.4 and figure 5.4A). This response has been interpreted as an example of evolutionary lag, i.e., the response was adaptive under conditions of intermittent food overabundance, as occurred over most of human history, but maladaptive under modern conditions of persistent food overabundance (Alcock et al., 2014). This interpretation may be true. However, we should consider the alternative hypothesis that some responses of the microbiome to environmental factors may persist because they are in the selective interests of the microorganisms, but not the host. In particular, a microorganism may respond adaptively to unfavorable conditions by increased investment in horizontal transmission, even where this is not in the selective interest of the host. This shift in strategy can be detected as a population bloom within the host, and resultant increase in the numbers of shed microbial cells. A likely example of such divergence of host and microbial interests is provided by certain horizontally transmitted bacteria in insects. Research on the black bean aphid *Aphis fabae* reveals that these bacteria have little or no impact on the fitness of their insect hosts when reared on nitrogen-sufficient plants, but display a massive population increase in hosts on low-nitrogen plants, depressing host fitness more than insects of the same genotype that lack these bacteria (Chandler et al., 2008). This and many other instances of context-dependent effects of individual microbial taxa or microbial communities on host phenotype and fitness illustrate how microbiome responses to environmental factors and to the interaction between host genotype and environment are not invariably to the advantage of the animal host.

The complexity of the multiway interactions and predicted diversity in the effects on host phenotype should be taken into account in the development of microbial therapies. Microbial communities and standardized microbial inocula may be beneficial for most people under most environmental conditions, but deleterious for certain genotypes under particular environmental circumstances. Thus, colonization of *Drosophila* with a standardized set of gut bacteria reduced lipid deposition of most fly genotypes, but had the reverse effect on a minority of genotypes (Dobson et al., 2015); and concerns about possible pro-obesogenic effects of probiotic bacteria added to many foods (Angelakis et al., 2013) should not be discounted.

8.3.3. THE INHERITANCE OF ACQUIRED CHARACTERISTICS

We have all learnt that traits acquired in an organism's lifetime are not inherited. Lamarck's giraffes that stretched their necks to reach food and produced offspring with longer necks have no place in modern biology founded on the neo-Darwinian synthesis.

We should not be so hasty on two counts, both of which involve the microbiome.

The first route for inheritance of acquired traits comes from the growing evidence that the phenotype of an animal can be influenced by epigenetic markers acquired during the lifetime of the parent and inherited faithfully to offspring (Noble, 2013). The biochemical basis of epigenesis is increasingly understood (Allis and Jenuwein, 2016), and several studies have implicated the microbiome. For example, in both mice and *Drosophila*, paternal exposure to high calorie diet can reprogram the chromatin markers in the sperm that determine availability of DNA for transcription, resulting in inherited obesity over subsequent generations (Carone et al., 2010; Ost et al., 2014). The many factors determining the genomic pattern of epigenetic markers are under intense investigation and, unsurprisingly, the microbiome is emerging as a significant correlation in some studies (Hullar and Fu, 2014; Kumar et al., 2014; Yu et al., 2015). These considerations raise the possibility of transgenerational microbiome effects mediating via epigenetic markers, as well as via mother-to-offspring transfer of microbial products discussed in section 8.1.3.

An alternative route for microbial-mediated inheritance of acquired characteristics relates to microorganisms with mixed horizontal and vertical modes of transmission. Microbial taxa with these traits are widespread in some invertebrates, notably various insects that transfer microbial partners vertically via the cytoplasm of the egg. An intensively studied system concerns the secondary symbiont bacteria *Hamiltonella defensa* and *Regiella insecticola*, which can protect their aphid hosts against parasitic wasps and fungal pathogens, respectively (Oliver et al., 2010). The selective advantage of these bacteria to the insect is, thus, strongly context-dependent and their prevalence in natural aphid populations is often very variable. Under conditions of high natural enemy abundance, aphids can acquire resistance by horizontal acquisition of these bacteria, and these traits can then be inherited with high fidelity by vertical transmission. The inheritance of these acquired traits can have profound effects on the fitness of aphid genotypes (McLean et al., 2016). More generally, the inheritance of acquired characteristics via microbial partners that have strongly context-dependent effects on host phenotype and mixed transmission modes may be a widespread determinant of animal phenotype. These effects would be largely overlooked in the many studies that focus on laboratory populations of animals with tightly controlled microbiology and also by both field and laboratory studies that quantify responses of host genotype frequencies to selection without any parallel consideration of microbiological changes.

8.4. The Animal in the Anthropocene

8.4.1. RESPONSES TO CLIMATE CHANGE

The impact of human activities on the planet has been formalized as a new geological era, the Anthropocene (Waters et al., 2016). As well as being characterized by various unintended consequences of human activities, including climate change, habitat fragmentation and loss, and biological extinctions, the Anthropocene brings the responsibility to manage and mitigate these effects on the health and well-being of humans and other organisms. Although rarely considered to date, the microbiome is an essential component of any full explanation or prediction of animal (including human) responses to anthropogenic factors.

We know that the microbiome of some animals varies with temperature. By far the most evident display of this effect comes from coral bleaching, where elevated sea surface temperatures cause the mass expulsion of symbiotic algae from reef-building corals, resulting in suppressed coral growth rates and, in extreme bleaching events, high coral mortality and consequent ecosystem changes (chapter 6, section 6.4.1). However, many other systems are affected by thermal stress. High temperatures can induce dramatic changes in the composition of the microbiota in marine invertebrates, sometimes associated with increases in the abundance of pathogens and susceptibility to disease (Lokmer and Mathias Wegner, 2015; Ritchie, 2006), and the microbial symbionts required by various insects are exquisitely sensitive to temperature (Wernegreen, 2012). Conversely, the microbiome of some animals appears to be robust to temperature fluctuations (Erwin et al., 2012), and certain microbial genotypes or taxa have been demonstrated to confer thermal tolerance in their animal hosts (Bay et al., 2016; Burke et al., 2010; Dunbar et al., 2007). This diversity of responses is important because it indicates that any consideration of animal responses to thermal stress associated with climate change should include a case-by-case consideration of the microbiome, including its response to temperature and the consequences of temperature-induced microbial changes for the health and fitness of the animal host. The environmental effects of climate change involve other environmental variables, including water availability in terrestrial habitats, salinity of some in-shore marine environments, and food quality. Consequently, the degree of animal cross-resistance to these different stressors, and their interaction with the microbiome are also important areas for future research.

The range of many animals is shifting in response to climate change, including both range expansions to higher latitudes and altitudes and contractions and local extinctions at the lower latitudinal and altitudinal limits. Many factors are recognized to influence the pattern and scale of these range shifts, e.g., the dispersal capabilities of the animals, direction of prevailing wind or current, and availability of suitable habitat (Chen et al., 2011; Moritz and Agudo, 2013). Should the properties of the microbiome also be considered? In principle, the microbiome can affect climate-related shifts in the range of animal species in two ways. First, the availability of suitable microbial partners in the new habitat may be crucial, particularly for open animal-microbial systems where animals exchange microbes with other taxa and free-living microbial communities. In this respect, data are almost entirely lacking, apart from some intriguing patterns relating to certain invasive animals, considered below (section 8.4.2). Second, microorganisms may influence the dispersal phenotype of animals. Phenotypic differences between animals at the advancing range margin and core of the range have been identified, mostly relating to dispersal capability, as influenced by patterns of energy metabolism and morphological features, such as wing length of insects (Kunin et al., 2009). A possible contribution of microorganisms to these phenotypic traits comes from evidence that microbial partners with mixed horizontal and vertical transmission can influence dispersal phenotype of some insects (Leonardo and Mondor, 2006) and energy metabolism traits of many animals (Nieuwdorp et al., 2014).

8.4.2. THE MICROBIAL DIMENSIONS OF INVASIVE ANIMALS

A very conspicuous consequence of human activities is invasive species, comprising animals or other taxa that are introduced, either on purpose or accidentally, to a location beyond their natural range, where they spread causing environmental and economic damage. The microbiome can influence the ecological success of introduced animal species in various ways, including

1. Microorganisms that are benign or beneficial for the introduced species may be pathogenic to native species
2. The introduced species may associate with microbial partners of native species, and the novel host-microbial combination may promote the invasiveness of the introduced species or have other deleterious effects
3. The introduced species may suppress microorganisms on which native taxa depend

To date, this area has received rather little attention, especially in comparison to the extensive research on microbial partners of invasive plants (Bunn et al., 2015), and the main focus has been the dynamics of microbial disease agents. There is abundant evidence that introduced animal species can be a reservoir for parasites and pathogens that have small or no effect on their fitness but are highly pathogenic to populations of native species. In essence, the microbial pathogens are the agents of apparent competition between the introduced and native animal taxa. Examples include the susceptibility of native noble crayfish in Europe to the fungal pathogen *Aphanomyces astaci* introduced with the signal crayfish from North America, and the deleterious effects of a microsporidian pathogen associated with the Asian harlequin ladybug beetle on native ladybugs in North America and Europe (Vilcinskas, 2015). The impact of microorganisms associated with introduced species can, additionally, extend beyond competitors of the introduced animal to other taxa, as is illustrated by the accidental introduction of the redbay ambrosia beetle, *Xyleborus glabratus* from Asia to the United States. As for other ambrosia beetles, adult *X. glabratus* beetles bore into the xylem vessels, usually of dead trees, to form galleries that are then inhabited by their larval offspring. The ambrosia beetles are associated with fungi, which line the galleries, providing food for the larvae, and are transmitted via specialized pockets (mycangia) in the exoskeleton of the adult insect. The invasive *X. glabratus* in the southeast United States has two intriguing features. First, it attacks live trees in its introduced range, even though it colonizes only dead trees in its native range (native U.S. ambrosia beetles also utilize dead trees only). Second, one of the ambrosial fungi that lines its galleries and is transmitted via mycangia is *Raffaelea lauricola*, a virulent pathogen of trees of the family

Lauraceae, including laurels and avocado. *R. lauricola*, transmitted to live trees via the introduced *X. glabratus*, is causing devastating losses of these plants (Hulcr and Stelinski, 2017).

Let us now consider the consequences of symbiont mixing between native and introduced animal species. This has been demonstrated in the association between wood wasps of the genus *Sirex* and their fungal symbionts. Following accidental introduction to North America, the European *Sirex noctilio* and its fungal partner *Amylostereum areolatum* are now sympatric with the native *S. nigricornis/A. chailletii* association. Some wasps of both species have switched fungal partners (Wooding et al., 2013), but how this shuffling of fungal partners influences the phenotype of the wasps remains to be determined. The consequences of an association between an introduced animal and native microorganism are more apparent for the scale insect *Cryptococcus fagisuga*, also introduced from Europe to North America. This insect feeds on beech trees, and its feeding behavior enables the otherwise nonpathogenic native fungus *Neonectria* to access the sieve elements of the tree, resulting in disease symptoms known as beech bark death and elevated tree mortality (Garnas et al., 2011). The relative taxonomic simplicity of the insect-fungal interactions and awareness of insect-vectored fungi as tree pathogens have facilitated identification of novel animal-microbial associations in these insect systems. Microbiological exchanges between native and introduced animals may occur in various other systems, and virtually nothing is known about the consequences for the phenotype and fitness of both the introduced and native species.

Introduced animal species may, additionally, suppress microorganisms on which native taxa depend. I know of no unambiguous example, but this type of interaction has some parallels to introduced species that alter the dynamics of pathogenic microorganisms in native species. For example, the introduction of the bank vole *Myodes* (=*Clethrionomys*) *glareolus* to Ireland has reduced the prevalence of the flea-transmitted *Bartonella* bacterium in wood mice *Apodemus sylvaticus*. This effect has been attributed to the low competence of bank voles as host for the fleas, so disrupting the transmission of *Bartonella* (Telfer et al., 2005). Equivalent scenarios involving mutualistic microorganisms can be envisaged. Consider a beneficial microorganism that is transmitted between individuals of a native animal species via the free-living environment, such that its persistence in the host population is dependent on sustained shedding from colonized hosts (see chapter 6, section 6.2 and figure 6.2B). Associations with these transmission traits are widespread in animals. If an introduced host species creates sink populations of the microorganism (i.e., the microorganism colonizes the host but is shed weakly) and is abundant, microbial transmission between its native hosts could be disrupted. For microorganisms that are beneficial to their native hosts, this interaction could, in principle, lead to a negative spiral of declining abundance of free-living populations of the microorganism and its native host, potentially facilitating the spread of the introduced species.

In summary, current understanding of the microbiological consequences of introduced animal species is shaped almost entirely by pathogens and disease. These effects are undoubtedly of major significance. However, we know that the health and fitness of animals are also shaped by their microbiome, which may be perturbed in multiple ways by the introduction of novel hosts with a microbiome of different function, composition or transmission dynamics. Microbiome research has the potential to contribute to efforts to explain and reduce the invasiveness of some introduced animal species.

8.4.3. MASS EXTINCTION AND THE MICROBIOME

In section 8.4.1, I addressed the susceptibility of some microbial associates of animals to thermal stress, a key feature of climate change, and in section 8.4.2 I described several routes by which invasive animals can, in principle, perturb the microbiome of native species with likely consequences for the phenotype and fitness of native taxa. These are not, however, the only routes by which the microbiome of animals can become compromised in the Anthropocene.

In relation to microbiome diversity loss, the greatest recent attention has been given to the human microbiome. There is evidence that the microbial diversity is reduced in human populations with Western lifestyles relative to hunter-gatherer lifestyles (Clemente et al., 2015; Schnorr et al., 2014); that *Helicobacter pylori*, a bacterial inhabitant of the stomach, has declined from very high prevalence to near extinction in some human populations over the last century (McJunkin et al., 2011); and that low microbial diversity can be associated with multiple chronic conditions, especially metabolic and immunological diseases, that have increased dramatically in recent years (Cox and Blaser, 2015). Proof-of-principle for a ratchet-like decline in microbial diversity has been obtained in a multigenerational study of mice fed on a diet that promotes simplification of the gut microbial community (chapter 3, section 3.5 and figure 3.13), although the relevance of this laboratory study to natural populations, including humans, is uncertain. Analyses of the microbiome of other animals are being driven by concerns over environmental pollution by antibiotics used in medicine and animal husbandry, as well as other antimicrobials (e.g., triclosan) used in consumer products (Dhillon et al., 2015; Martinez, 2009). The continued research efforts on the impacts of these products on the microbiome of natural animal populations will both benefit from and contribute to greater understanding of the effects of these products on the human microbiome.

As discussed in chapter 3 (section 3.5.3), probiotics and prebiotics to promote the diversity of the human gut microbiome are seen as strategies for restoration of microbiomes that have been depleted by antimicrobials, excessive cleanliness, and inappropriate diets. These approaches, however, presuppose a high degree of functional redundancy with low incidence of taxa specifically adapted to certain host genotypes and weak coevolutionary interactions between different members of the microbiota and the host. As discussed in chapters 6 and 7, our understanding of the ecological and evolutionary interactions in these complex systems is very limited. There is an urgent need for interdisciplinary research to establish the feasibility and most effective routes for ecological restoration of the microbiome from both the biomedical and environmental perspectives.

As these considerations illustrate, microbiome research combines the study of genuinely complex and difficult fundamental questions with the need to solve pressing problems of biomedical and environmental concern. The solutions to these problems require two conceptual realignments: to recognize the central importance of the microbiology of animals, and to recognize their study is an interdisciplinary endeavor.

REFERENCES

Aagaard, K., Ma, J., Antony, K.M., Ganu, R., Petrosino, J., and Versalovic, J. (2014). The placenta harbors a unique microbiome. *Sci Transl Med* 6, 237ra265.

Acuna, R., Padilla, B.E., Florez-Ramos, C.P., Rubio, J.D., Herrera, J.C., Benavides, P., Lee, S.J., et al. (2012). Adaptive horizontal transfer of a bacterial gene to an invasive insect pest of coffee. *Proc Natl Acad Sci USA* 109, 4197–202.

Alcock, J., Maley, C.C., and Aktipis, C.A. (2014). Is eating behavior manipulated by the gastrointestinal microbiota? Evolutionary pressures and potential mechanisms. *Bioessays* 36, 940–49.

Alegado, R.A., Brown, L.W., Cao, S., Dermenjian, R.K., Zuzow, R., Fairclough, S.R., Clardy, J., and King, N. (2012). A bacterial sulfonolipid triggers multicellular development in the closest living relatives of animals. *Elife* 1, e00013.

Allis, C.D., and Jenuwein, T. (2016). The molecular hallmarks of epigenetic control. *Nat Rev Genet* 17, 487–500.

Almond, D., and Currie, J. (2011). Killing me softly: the fetal origins hypothesis. *J Econ Perspect* 25, 153–72.

Aminov, R.I. (2011). Horizontal gene exchange in environmental microbiota. *Front Microbiol* 2, 158.

Amir, I., Konikoff, F.M., Oppenheim, M., Gophna, U., and Half, E.E. (2014). Gastric microbiota is altered in oesophagitis and Barrett's oesophagus and further modified by proton pump inhibitors. *Environ Microbiol* 16, 2905–14.

Angelakis, E., Merhej, V., and Raoult, D. (2013). Related actions of probiotics and antibiotics on gut microbiota and weight modification. *Lancet Infect Dis* 13, 889–99.

Arumugam, M., Raes, J., Pelletier, E., Le Paslier, D., Yamada, T., Mende, D.R., Fernandes, G.R., et al. (2011). Enterotypes of the human gut microbiome. *Nature* 473, 174–80.

Ayres, J.S., and Schneider, D.S. (2012). Tolerance of infections. *Annu Rev Immunol* 30, 271–94

Bai, Y., Muller, D.B., Srinivas, G., Garrido-Oter, R., Potthoff, E., Rott, M., Dombrowski, N., et al. (2015). Functional overlap of the Arabidopsis leaf and root microbiota. *Nature* 528, 364–69.

Barker, D.J.P. (1992). *Fetal and Infant Origins of Adult Disease* (London, UK: BMJ Books).

Barr, J.J., Auro, R., Furlan, M., Whiteson, K.L., Erb, M.L., Pogliano, J., Stotland, A., et al. (2013). Bacteriophage adhering to mucus provide a non-host-derived immunity. *Proc Natl Acad Sci USA* 110, 10771–76.

Bashir, M.E., Louie, S., Shi, H.N., and Nagler-Anderson, C. (2004). Toll-like receptor 4 signaling by intestinal microbes influences susceptibility to food allergy. *J Immunol* 172, 6978–87.

Bates, J.M., Mittge, E., Kuhlman, J., Baden, K.N., Cheesman, S.E., and Guillemin, K. (2006). Distinct signals from the microbiota promote different aspects of zebrafish gut differentiation. *Dev Biol* 297, 374–86.

Baxter, N.T., Wan, J.J., Schubert, A.M., Jenior, M.L., Myers, P., and Schloss, P.D. (2015). Intra- and interindividual variations mask interspecies variation in the microbiota of sympatric peromyscus populations. *Appl Environ Microbiol* 81, 396–404.

Bay, L.K., Doyle, J., Logan, M., and Berkelmans, R. (2016). Recovery from bleaching is medi-ated by threshold densities of background thermo-tolerant symbiont types in a reef-building coral. *Roy Soc Open Sci* 3, 160322.

Bercik, P., Denou, E., Collins, J., Jackson, W., Lu, J., Jury, J., Deng, Y., et al. (2011). The intestinal microbiota affect central levels of brain-derived neurotropic factor and behavior in mice. *Gastroenterol* 141, 599–609.

Berg, M., Stenuit, B., Ho, J., Wang, A., Parke, C., Knight, M., Alvarez-Cohen, L., and Shapira, M. (2016). Assembly of the *Caenorhabditis elegans* gut microbiota from diverse soil microbial environments. *ISME J* 10, 1998–2009.

Berkelmans, R., and van Oppen, M.J. (2006). The role of zooxanthellae in the thermal tolerance of corals: a 'nugget of hope' for coral reefs in an era of climate change. *Proc Biol Sci* 273, 2305–12.

Berry, D., and Widder, S. (2014). Deciphering microbial interactions and detecting keystone species with co-occurrence networks. *Front Microbiol* 5, 219.

Bever, J.D., and Simms, E.L. (2000). Evolution of nitrogen fixation in spatially structured popu-lations of *Rhizobium*. *Heredity* 85, 366–72.

Bhargava, P., and Mowry, E.M. (2014). Gut microbiome and multiple sclerosis. *Curr Neurol Neurosci Rep* 14, 492.

Bik, E.M., Eckburg, P.B., Gill, S.R., Nelson, K.E., Purdom, E.A., Francois, F., Perez-Perez, G., Blaser, M.J., and Relman, D.A. (2006). Molecular analysis of the bacterial microbiota in the human stomach. *Proc Natl Acad Sci USA* 103, 732–37.

Blaser, M.J., and Falkow, S. (2009). What are the consequences of the disappearing human mi-crobiota? *Nat Rev Microbiol* 7, 887–94.

Bonfante, P., and Anca, I.A. (2009). Plants, mycorrhizal fungi, and bacteria: a network of inter-actions. *Annu Rev Microbiol* 63, 363–83.

Booijink, C.C., El-Aidy, S., Rajilic-Stojanovic, M., Heilig, H.G., Troost, F.J., Smidt, H., Kleere-bezem, M., De Vos, W.M., and Zoetendal, E.G. (2010). High temporal and inter-individual variation detected in the human ileal microbiota. *Environ Microbiol* 12, 3213–27.

Boon, E., Meehan, C.J., Whidden, C., Wong, D.H., Langille, M.G., and Beiko, R.G. (2014). In-teractions in the microbiome: communities of organisms and communities of genes. *FEMS Microbiol Rev* 38, 90–118.

Bravo, J.A., Forsythe, P., Chew, M.V., Escaravage, E., Savignac, H.M., Dinan, T.G., Bienenstock, J., and Cryan, J.F. (2011). Ingestion of *Lactobacillus* strain regulates emotional behavior and central GABA receptor expression in a mouse via the vagus nerve. *Proc Natl Acad Sci USA* 108, 16050–55.

Breeuwer, J.A.J., and Werren, J.H. (1995). Hybrid breakdown between two haplodiploid spe-cies: the role of nuclear and cytoplasmic genes. *Evolution* 49, 705–17.

Brock, D.A., Douglas, T.E., Queller, D.C., and Strassmann, J.E. (2011). Primitive agriculture in a social amoeba. *Nature* 469, 393–96.

Broderick, N.A., Buchon, N., and Lemaitre, B. (2014). Microbiota-induced changes in *Drosoph-ila melanogaster* host gene expression and gut morphology. *mBio* 5, e01117–14.

Brown, B.E. (1997). Coral bleaching: causes and consequences. *Coral Reefs* 16, S129–38.

Brown, B.E., Dunne, R.P., Goodson, M.S., and Douglas, A.E. (2000). Bleaching patterns in reef corals. *Nature* 404, 142–43.

Brucker, R.M., and Bordenstein, S.R. (2012). Speciation by symbiosis. *Trends Ecol Evol* 27, 443–51.

Brucker, R.M., and Bordenstein, S.R. (2013). The hologenomic basis of speciation: gut bacteria cause hybrid lethality in the genus *Nasonia*. *Science* 341, 667–69.

Buchner, P. (1965). *Endosymbioses of Animals with Plant Micro-Organisms* (Chichester, UK: John Wiley and Sons).

Buchon, N., Broderick, N.A., Chakrabarti, S., and Lemaitre, B. (2009). Invasive and indigenous microbiota impact intestinal stem cell activity through multiple pathways in *Drosophila*. *Genes Dev* 23, 2333–44.

Buffie, C.G., Jarchum, I., Equinda, M., Lipuma, L., Gobourne, A., Viale, A., Ubeda, C., Xavier, J., and Pamer, E.G. (2012). Profound alterations of intestinal microbiota following a single dose of clindamycin results in sustained susceptibility to *Clostridium difficile*-induced colitis. *Infect Immun* 80, 62–73.

Buffie, C.G., Bucci, V., Stein, R.R., McKenney, P.T., Ling, L., Gobourne, A., No, D., et al. (2015). Precision microbiome reconstitution restores bile acid mediated resistance to *Clostridium difficile*. *Nature* 517, 205–8.

Bunn, R.A., Ramsey, P.W., and Lekberg, Y. (2015). Do native and invasive plants differ in their interactions with arbuscual mycorrhizal fungi? A meta-analysis. *J Ecol* 103, 1547–56.

Buonaurio, R., Moretti, C., da Silva, D.P., Cortese, C., Ramos, C., and Venturi, V. (2015). The olive knot disease as a model to study the role of interspecies bacterial communities in plant disease. *Front Plant Sci* 6, 434.

Burke, G., Fiehn, O., and Moran, N. (2010). Effects of facultative symbionts and heat stress on the metabolome of pea aphids. *ISME J* 4, 242–52.

Burki, F. (2014). The eukaryotic tree of life from a global phylogenomic perspective. *Cold Spring Harb Perspect Biol* 6, a016147.

Campbell, M.A., Van Leuven, J.T., Meister, R.C., Carey, K.M., Simon, C., and McCutcheon, J.P. (2015). Genome expansion via lineage splitting and genome reduction in the cicada endo-symbiont *Hodgkinia*. *Proc Natl Acad Sci USA* 112, 10192–99.

Calderon-Cortes, N., Quesada, M., Watanabe, H., Cano-Camacho, H., and Oyama, K. (2012). Endogenous plant cell wall digestion: a key mechanism in insect evolution. *Annu Rev Ecol Evol Syst* 43, 45–71.

Candela, M., Biagi, E., Maccaferri, S., Turroni, S., and Brigidi, P. (2012). Intestinal microbiota is a plastic factor responding to environmental changes. *Trends Microbiol* 20, 385–91.

Cani, P.D., Amar, J., Iglesias, M.A., Poggi, M., Knauf, C., Bastelica, D., Neyrinck, A.M., et al. (2007). Metabolic endotoxemia initiates obesity and insulin resistance. *Diabetes* 56, 1761–72.

Caporaso, J.G., Lauber, C.L., Costello, E.K., Berg-Lyons, D., Gonzalez, A., Stombaugh, J., Knights, D., et al. (2011). Moving pictures of the human microbiome. *Genome Biol* 12, R50.

Caragata, E.P., Rancès, E., Hedges, L.M., Gofton, A.W., Johnson, K.N., O'Neill, S.L., and Mc-Graw, E.A. (2013). Dietary cholesterol modulates pathogen blocking by *Wolbachia*. *PLoS Pathog* 9, e1003459.

Caricilli, A.M., Picardi, P.K., de Abreu, L.L., Ueno, M., Prada, P.O., Ropelle, E.R., Hirabara, S.M., et al. (2011). Gut microbiota is a key modulator of insulin resistance in TLR 2 knockout mice. *PLoS Biol* 9, e1001212.

Carone, B.R., Fauquier, L., Habib, N., Shea, J.M., Hart, C.E., Li, R., Bock, C., et al. (2010). Paternally induced transgenerational environmental reprogramming of metabolic gene expression in mammals. *Cell* 143, 1084–96.

Carro, A.C., and Damonte, E.B. (2013). Requirement of cholesterol in the viral envelope for dengue virus infection. *Virus Res* 174, 78–87

Ceja-Navarro, J.A., Vega, F.E., Karaoz, U., Hao, Z., Jenkins, S., Lim, H.C., Kosina, P., Infante, F., Northen, T.R., and Brodie, E.L. (2015). Gut microbiota mediate caffeine detoxification in the primary insect pest of coffee. *Nat Commun* 6, 7618.

Chandler, J.A., Eisen, J.A., and Kopp, A. (2012). Yeast communities of diverse *Drosophila* species: comparison of two symbiont groups in the same hosts. *Appl Environ Microbiol* 78, 7327–36.

Chandler, J.A., and Turelli, M. (2014). Comment on "The hologenomic basis of speciation: gut bacteria cause hybrid lethality in the genus *Nasonia*." *Science* 345, 1011.

Chandler, S.M., Wilkinson, T.L., and Douglas, A.E. (2008). Impact of plant nutrients on the relationship between a herbivorous insect and its symbiotic bacteria. *Proc Biol Sci* 275, 565–70.

Chang, K.P., Chang, C.S., and Sassa, S. (1975). Heme biosynthesis in bacterium-protozoon symbioses: enzymic defects in host hemoflagellates and complemental role of their intracellular symbiotes. *Proc Natl Acad Sci USA* 72, 2979–83.

Charbonneau, M.R., Blanton, L.V., DiGiulio, D.B., Relman, D.A., Lebrilla, C.B., Mills, D.A., and Gordon, J.I. (2016). A microbial perspective of human developmental biology. *Nature* 535, 48–55.

Chaston, J.M., Dobson, A.J., Newell, P.D., and Douglas, A.E. (2016). Host genetic control of the microbiota mediates the *Drosophila* nutritional phenotype. *Appl Environ Microbiol* 82, 671–79.

Chen, I.C., Hill, J.K., Ohlemuller, R., Roy, D.B., and Thomas, C.D. (2011). Rapid range shifts of species associated with high levels of climate warming. *Science* 333, 1024–26.

Chevalier, C., Stojanovic, O., Colin, D.J., Suarez-Zamorano, N., Tarallo, V., Veyrat-Durebex, C., Rigo, D., et al. (2015). Gut microbiota orchestrates energy homeostasis during cold. *Cell* 163, 1360–74.

Clarke, T.B., Davis, K.M., Lysenko, E.S., Zhou, A.Y., Yu, Y., and Weiser, J.N. (2010). Recognition of peptidoglycan from the microbiota by Nod1 enhances systemic innate immunity. *Nat Med* 16, 228–31.

Clemente, J.C., Pehrsson, E.C., Blaser, M.J., Sandhu, K., Gao, Z., Wang, B., Magris, M., et al. (2015). The microbiome of uncontacted Amerindians. *Sci Adv* 1, e1500183.

Cohen, J. (2016). Vaginal microbiome affects HIV risk. *Science* 353, 331.

Costello, E.K., Stagaman, K., Dethlefsen, L., Bohannan, B.J., and Relman, D.A. (2012). The application of ecological theory toward an understanding of the human microbiome. *Science* 336, 1255–62.

Cox, L.M., and Blaser, M.J. (2015). Antibiotics in early life and obesity. *Nat Rev Endocrinol* 11, 182–90.

Cox, L.M., Yamanishi, S., Sohn, J., Alekseyenko, A.V., Leung, J.M., Cho, I., Kim, S.G., et al. (2014). Altering the intestinal microbiota during a critical developmental window has lasting metabolic consequences. *Cell* 158, 705–21.

Coyte, K.Z., Schluter, J., and Foster, K.R. (2015). The ecology of the microbiome: networks, competition, and stability. *Science* 350, 663–66.

Crippen, T.L., and Poole, T.L. (2009). Conjugative transfer of plasmid-located antibiotic resistance genes within the gastrointestinal tract of lesser mealworm larvae, *Alphitobius diaperinus* (Coleoptera: Tenebrionidae). *Foodborne Pathog* Dis 6, 907–15.

Cullender, T.C., Chassaing, B., Janzon, A., Kumar, K., Muller, C.E., Werner, J.J., Angenent, L.T., et al. (2013). Innate and adaptive immunity interact to quench microbiome flagellar motility in the gut. *Cell Host Microbe* 14, 571–81.

Daane, L.L., Molina, J., and Sadowsky, M.J. (1997). Plasmid transfer between spatially separated donor and recipient bacteria in earthworm-containing soil microcosms. *Appl Environ Microbiol* 63, 679–86.

Dantzer, R., O'Connor, J.C., Freund, G.G., Johnson, R.W., and Kelley, K.W. (2008). From inflammation to sickness and depression: when the immune system subjugates the brain. *Nat Rev Neurosci* 9, 46–56.

David, L.A., Materna, A.C., Friedman, J., Campos-Baptista, M.I., Blackburn, M.C., Perrotta, A., Erdman, S.E., and Alm, E.J. (2014). Host lifestyle affects human microbiota on daily timescales. *Genome Biol* 15, R89.

David, L.A., Maurice, C.F., Carmody, R.N., Gootenberg, D.B., Button, J.E., Wolfe, B.E., Ling, A.V., et al. (2014). Diet rapidly and reproducibly alters the human gut microbiome. *Nature* 505, 559–63.

Davidson, S.K., Koropatnick, T.A., Kossmehl, R., Sycuro, L., McFall-Ngai, M.J. (2004). NO means 'yes' in the squid-vibrio symbiosis: nitric oxide (NO) during the initial stages of a beneficial association. *Cell Microbiol* 6, 1139–51.

Davies, J., and Ryan, K.S. (2012). Introducing the parvome: bioactive compounds in the microbial world. *ACS Chem Biol* 7, 252–59.

Davis, M.P., Sparks, J.S., and Smith, W.L. (2016). Repeated and widespread evolution of bioluminescence in marine fishes. *PLoS One* 11, e0155154.

De Filippo, C., Cavalieri, D., Di Paola, M., Ramazzotti, M., Poullet, J.B., Massart, S., Collini, S., Pieraccini, G., and Lionetti, P. (2010). Impact of diet in shaping gut microbiota revealed by a comparative study in children from Europe and rural Africa. *Proc Natl Acad Sci USA* 107, 14691–96.

Debast, S.B., Bauer, M.P., and Kuijper, E.J. (2014). Update of the treatment guidance document for *Clostridium difficile* infection. *Clin Microbiol Infect* 20 Suppl 2, 1–26.

Degnan, P.H., Pusey, A.E., Lonsdorf, E.V., Goodall, J., Wroblewski, E.E., Wilson, M.L., Rudicell, R.S., Hahn, B.H., and Ochman, H. (2012). Factors associated with the diversification of the gut microbial communities within chimpanzees from Gombe National Park. *Proc Natl Acad Sci USA* 109, 13034–39.

Degnan, S.M. (2015). The surprisingly complex immune gene repertoire of a simple sponge, exemplified by the NLR genes: a capacity for specificity? *Dev Comp Immunol* 48, 269–74.

Denison, R.F., and Kiers, E.T. (2011). Life histories of symbiotic rhizobia and mycorrhizal fungi. *Curr Biol* 21, R775–85.

Desbonnet, L., Clarke, G., Shanahan, F., Dinan, T.G., and Cryan, J.F. (2014). Microbiota is essential for social development in the mouse. *Mol Psychiatry* 19, 146–48.

Dethlefsen, L., McFall-Ngai, M., and Relman, D.A. (2007). An ecological and evolutionary perspective on human-microbe mutualism and disease. *Nature* 449, 811–18.

Dethlefsen, L., and Relman, D.A. (2011). Incomplete recovery and individualized responses of the human distal gut microbiota to repeated antibiotic perturbation. *Proc Natl Acad Sci USA* 108 Suppl 1, 4554–61.

Dewhirst, F.E., Chien, C.C., Paster, B.J., Ericson, R.L., Orcutt, R.P., Schauer, D.B., and Fox, J.G. (1999). Phylogeny of the defined murine microbiota: Altered Schaedler Flora. *Appl Environ Microbiol* 65, 3287–92.

Dhillon, G.S., Kaur, S., Pulicharla, R., Brar, S.K., Cledon, M., Verma, M., and Surampalli, R.Y. (2015). Triclosan: current status, occurrence, environmental risks and bioaccumulation potential. *Int J Environ Res Public Health* 12, 5657–84.

Dickson, R.P., Erb-Downward, J.R., Martinez, F.J., and Huffnagle, G.B. (2016). The microbiome and the respiratory tract. *Annu Rev Physiol* 78, 481–504.

Dimitriu, T., Lotton, C., Benard-Capelle, J., Misevic, D., Brown, S.P., Lindner, A.B., and Taddei, F. (2014). Genetic information transfer promotes cooperation in bacteria. *Proc Natl Acad Sci USA* 111, 11103–8.

Dobson, A.J., Chaston, J.M., Newell, P.D., Donahue, L., Hermann, S.L., Sannino, D.R., Westmiller, S., et al. (2015). Host genetic determinants of microbiota-dependent nutrition revealed by genome-wide analysis of *Drosophila melanogaster*. *Nat Commun* 6, 6312.

Dominguez-Bello, M.G., Costello, E.K., Contreras, M., Magris, M., Hidalgo, G., Fierer, N., and Knight, R. (2010). Delivery mode shapes the acquisition and structure of the initial microbiota across multiple body habitats in newborns. *Proc Natl Acad Sci USA* 107, 11971–75.

Donia, M.S., Hathaway, B.J., Sudek, S., Haygood, M.G., Rosovitz, M.J., Ravel, J., and Schmidt, E.W. (2006). Natural combinatorial peptide libraries in cyanobacterial symbionts of marine ascidians. *Nature Chem Biol* 2, 729–35

Douglas, A.E. (1988). Experimental studies on the mycetome symbiosis in the leafhopper *Euscelis incisus*. *J Insect Physiol* 34, 1043–53.

Douglas, A.E. (2003). Coral bleaching—how and why? *Marine Pollution Bulletin* 46, 385–92.

Douglas, A.E. (2010). *The Symbiotic Habit* (Princeton, NJ: Princeton University Press).

Douglas, A.E. (2015). Multiorganismal insects: diversity and function of resident microorganisms. *Annu Rev Entomol* 60, 17–34.

Douglas, A.E. (2016). How multi-partner endosymbioses function. *Nature Rev Microbiol* 14, 731–43.

Douglas, A.E., and Dobson, A.J. (2013). New synthesis: animal communication mediated by microbes: fact or fantasy? *J Chem Ecol* 39, 1149.

Douglas, A.E., and Smith, D.C. (1983). The cost of symbionts to their host in green hydra. *Endocytobiol* 2, 631–48.

Douglas, A.E., and Werren, J.H. (2016). Holes in the hologenome: why host-microbe symbioses are not holobionts. *mBio* 7, e02099.

Douglas, G.L., and Klaenhammer, T.R. (2010). Genomic evolution of domesticated microorganisms. *Annu Rev Food Sci Technol* 1, 397–414.

Drissi, F., Merhej, V., Angelakis, E., El Kaoutari, A., Carriere, F., Henrissat, B., and Raoult, D. (2014). Comparative genomics analysis of *Lactobacillus* species associated with weight gain or weight protection. *Nutr Diabetes* 4, e109.

Dubilier, N., Bergin, C., and Lott, C. (2008). Symbiotic diversity in marine animals: the art of harnessing chemosynthesis. *Nature Rev Microbiol* 6, 725–40.

Dunbar, H.E., Wilson, A.C., Ferguson, N.R., and Moran, N.A. (2007). Aphid thermal tolerance is governed by a point mutation in bacterial symbionts. *PLoS Biol* 5, e96.

Dunning Hotopp, J.C., Clark, M.E., Oliveira, D.C., Foster, J.M., Fischer, P., Munoz Torres, M.C., Giebel, J.D., et al. (2007). Widespread lateral gene transfer from intracellular bacteria to multicellular eukaryotes. *Science* 317, 1753–56.

Elliot, S.L., Blanford, S., and Thomas, M.B. (2002). Host-pathogen interactions in a varying environment: temperature, behavioural fever and fitness. *Proc Biol Sci* 269, 1599–607.

Embley, T.M., van der Giezen, M., Horner, D.S., Dyal, P.L., and Foster, P. (2003). Mitochondria and hydrogenosomes are two forms of the same fundamental organelle. *Philos Trans R Soc Lond B* 358, 191–201.

Erejuwa, O.O., Sulaiman, S.A., and Ab Wahab, M.S. (2014). Modulation of gut microbiota in the management of metabolic disorders: the prospects and challenges. *Int J Mol Sci* 15, 4158–88.

Erwin, P.M., Pita, L., Lopez-Legentil, S., and Turon, X. (2012). Stability of sponge-associated bacteria over large seasonal shifts in temperature and irradiance. *Appl Environ Microbiol* 78, 7358–68.

Everard, A., Belzer, C., Geurts, L., Ouwerkert, J.P, Druart, C., Bindels, L.B., Guiot, Y., et al. (2013). Cross-talk between *Akkermansia muciniphila* and intestinal epithelium controls diet-induced obesity. *Proc Natl Acad. Sci USA* 110, 9066–71.

Ezenwa, V.O., Archie, E.A., Craft, M.E., Hawley, D.M., Martin, L.B., Moore, J., and White, L. (2016). Host behaviour-parasite feedback: an essential link between animal behaviour and disease ecology. *Proc Biol Sci* 283, 20153078.

Ezenwa, V.O., and Williams, A.E. (2014). Microbes and animal olfactory communication: where do we go from here? *Bioessays* 36, 847–54.

Faith, J.J., Guruge, J.L., Charbonneau, M., Subramanian, S., Seedorf, H., Goodman, A.L., Clemente, J.C., et al. (2013). The long-term stability of the human gut microbiota. *Science* 341, 1237439.

Faust, K., Lahti, L., Gonze, D., de Vos, W.M., and Raes, J. (2015). Metagenomics meets time series analysis: unraveling microbial community dynamics. *Curr Opin Microbiol* 25, 56–66.

Faust, K., and Raes, J. (2012). Microbial interactions: from networks to models. *Nature Rev Microbiol* 10, 538–50.

Faust, K., and Raes, J. (2016). Host-microbe interaction: Rules of the game for microbiota. *Nature* 534, 182–83.

Faust, K., Sathirapongsasuti, J.F., Izard, J., Segata, N., Gevers, D., Raes, J., and Huttenhower, C. (2012). Microbial co-occurrence relationships in the human microbiome. *PLoS Comput Biol* 8, e1002606.

Felix, M.A., and Duveau, F. (2012). Population dynamics and habitat sharing of natural populations of *Caenorhabditis elegans* and *C. briggsae*. *BMC Biol* 10, 59.

Fenchel, T., and Finlay, B.J. (1995). *Ecology and Evolution in Anoxic Worlds* (Oxford, UK: Oxford University Press).

Fenton, A., and Perkins, S.E. (2010). Applying predator-prey theory to modelling immune-mediated, within-host interspecific parasite interactions. *Parasitol* 137, 1027–38.

Fetissov, S.O., Hamze Sinno, M., Coeffier, M., Bole-Feysot, C., Ducrotte, P., Hokfelt, T., and Dechelotte, P. (2008). Autoantibodies against appetite-regulating peptide hormones and neuropeptides: putative modulation by gut microflora. *Nutrition* 24, 348–359.

Florez, L.V., Biedermann, P.H., Engl, T., and Kaltenpoth, M. (2015). Defensive symbioses of animals with prokaryotic and eukaryotic microorganisms. *Natural Prod Rep* 32, 904–36.

Forsythe, P., Kunze, W., and Bienenstock, J. (2016). Moody microbes or fecal phrenology: what do we know about the microbiota-gut-brain axis? *BMC Med* 14, 58.

Franzenburg, S., Fraune, S., Altrock, P.M., Kunzel, S., Baines, J.F., Traulsen, A., and Bosch, T.C. (2013). Bacterial colonization of Hydra hatchlings follows a robust temporal pattern. *ISME J* 7, 781–90.

Franzenburg, S., Walter, J., Kunzel, S., Wang, J., Baines, J.F., Bosch, T.C., and Fraune, S. (2013). Distinct antimicrobial peptide expression determines host species-specific bacterial associations. *Proc Natl Acad Sci USA* 110, E3730–38.

Franzosa, E.A., Morgan, X.C., Segata, N., Waldron, L., Reyes, J., Earl, A.M., Giannoukos, G., et al. (2014). Relating the metatranscriptome and metagenome of the human gut. *Proc Natl Acad Sci USA* 111, E2329–38.

Frases, S., Chaskes, S., Dadachova, E., and Casadevall, A. (2006). Induction by *Klebsiella aerogenes* of a melanin-like pigment in *Cryptococcus neoformans*. *Appl Environ Microbiol* 72, 1542–50.

Fraune, S., and Bosch, T.C. (2007). Long-term maintenance of species-specific bacterial microbiota in the basal metazoan *Hydra*. *Proc Natl Acad Sci USA* 104, 13146–51.

Fraune, S., Anton-Erxleben, F., Augustin, R., Franzenburg, S., Knop, M., Schröder, K., Willoweit-Ohl, D., and Bosch, T.C. (2015). Bacteria-bacteria interactions within the microbiota of the ancestral metazoan *Hydra* contribute to fungal resistance. *ISME J* 9, 1543–56.

Frese, S.A., Benson, A.K., Tannock, G.W., Loach, D.M., Kim, J., Zhang, M., Oh, P.L., et al. (2011). The evolution of host specialization in the vertebrate gut symbiont *Lactobacillus reuteri*. *PLoS Genet* 7, e1001314.

Frost, G., Sleeth, M.L., Sahuri-Arisoylu, M., Lizarbe, B., Cerdan, S., Brody, L., Anastasovska, J., et al. (2014). The short-chain fatty acid acetate reduces appetite via a central homeostatic mechanism. *Nat Commun* 5, 3611.

Fukatsu, T., and Hosokawa, T. (2002). Capsule-transmitted gut symbiotic bacterium of the Japanese common plataspid stinkbug, *Megacopta punctatissima*. *Appl Environ Microbiol* 68, 389–96.

Gajer, P., Brotman, R.M., Bai, G., Sakamoto, J., Schutte, U.M., Zhong, X., Koenig, S.S., et al. (2012). Temporal dynamics of the human vaginal microbiota. *Sci Transl Med* 4, 132ra152.

Garnas, J.R., Ayres, M.P., Liebhold, A.M., and Evans, C. (2011). Subcontinental impacts of an invasive tree disease on forest structure and dynamics. *J Ecol* 99, 532–41.

Gendrin, M., Rodgers, F.H., Yerbanga, R.S., Ouedraogo, J.B., Basanez, M.G., Cohuet, A., and Christophides, G.K. (2015). Antibiotics in ingested human blood affect the mosquito microbiota and capacity to transmit malaria. *Nat Commun* 6, 5921.

Gerardo, N.M., and Parker, B.J. (2014). Mechanisms of symbiont-conferred protection against natural enemies: an ecological and evolutionary framework. *Curr Op Insect Sci* 4, 8–14.

Ghoul, M., Griffin, A.S., and West, S.A. (2014). Toward an evolutionary definition of cheating. *Evolution* 68, 318–31.

Gil-Turnes, M.S., Hay, M.E., and Fenical, W. (1989). Symbiotic marine bacteria chemically defend crustacean embryos from a pathogenic fungus. *Science* 246, 116–18

Goffredi, S.K., Gregory, A., Jones, W.J., Morella, N.M., and Sakamoto, R.I. (2014). Ontogenetic variation in epibiont community structure in the deep-sea yeti crab, *Kiwa puravida*: convergence among crustaceans. *Mol Ecol* 23, 1457–72.

Goodrich, J.K., Waters, J.L., Poole, A.C., Sutter, J.L., Koren, O., Blekhman, R., Beaumont, M., et al. (2014). Human genetics shape the gut microbiome. *Cell* 159, 789–99.

Gorman, M.L. (1976). A mechanism for individual recognition by odour in *Herpestes auropunctatus* (Carnivora: Viverridae). *Anim Behav* 24, 141–45.

Goto, Y., Obata, T., Kunisawa, J., Sato, S., Ivanov, I.I., Lamichhane, A., Takeyama, N., et al. (2014). Innate lymphoid cells regulate intestinal epithelial cell glycosylation. *Science* 345, 1254009.

Hamady, M., and Knight, R. (2009). Microbial community profiling for human microbiome projects: Tools, techniques, and challenges. *Genome Res* 19, 1141–52.

Hamilton, P.T., Peng, F., Boulanger, M.J., and Perlman, S.J. (2016). A ribosome-inactivating protein in a *Drosophila* defensive symbiont. *Proc Natl Acad Sci USA* 113, 350–55

Hansen, A.K., and Moran, N.A. (2014). The impact of microbial symbionts on host plant utilization by herbivorous insects. *Mol Ecol* 23, 1473–96.

Hart, B.L. (1994). Behavioural defense against parasites: interaction with parasite invasiveness. *Parasitol* 109 Suppl, S139–151.

Heath-Heckman, E.A., Peyer, S.M., Whistler, C.A., Apicella, M.A., Goldman, W.E., and McFall-Ngai, M.J. (2013). Bacterial bioluminescence regulates expression of a host cryptochrome gene in the squid-*Vibrio* symbiosis. *mBio* 4, e00167–13.

Hedges, L.M., Brownlie, J.C., O'Neill, S.L., and Johnson, K.N. (2008). *Wolbachia* and virus protection in insects. *Science* 322, 702.

Hehemann, J.H., Correc, G., Barbeyron, T., Helbert, W., Czjzek, M., and Michel, G. (2010). Transfer of carbohydrate-active enzymes from marine bacteria to Japanese gut microbiota. *Nature* 464, 908–12.

Hehemann, J.H., Kelly, A.G., Pudlo, N.A., Martens, E.C., and Boraston, A.B. (2012). Bacteria of the human gut microbiome catabolize red seaweed glycans with carbohydrate-active enzyme updates from extrinsic microbes. *Proc Natl Acad Sci USA* 109, 19786–91.

Henry, L.M., Peccoud, J., Simon, J.C., Hadfield, J.D., Maiden, M.J., Ferrari, J., and Godfray, H.C. (2013). Horizontally transmitted symbionts and host colonization of ecological niches. *Curr Biol* 23, 1713–17.

Hentschel, U., Piel, J., Degnan, S.M., and Taylor, M.W. (2012). Genomic insights into the marine sponge microbiome. *Nat Rev Microbiol* 10, 641–54.

Hill, C., Guarner, F., Reid, G., Gibson, G.R., Merenstein, D.J., Pot, B., Morelli, L., et al. (2014). The International Scientific Association for Probiotics and Prebiotics consensus statement on the scope and appropriate use of the term probiotic. *Nature Rev Gastroent Hepat* 11, 506–14.

Hittinger, C.T. (2013). *Saccharomyces* diversity and evolution: a budding model genus. *Trends Genet* 29, 309–17.

Hogan, D.A., Vik, A., and Kolter, R. (2004). A *Pseudomonas aeruginosa* quorum-sensing molecule influences *Candida albicans* morphology. *Mol Microbiol* 54, 1212–23.

Hold, G.L., Smith, M., Grange, C., Watt, E.R., El-Omar, E.M., and Mukhopadhya, I. (2014). Role of the gut microbiota in inflammatory bowel disease pathogenesis: what have we learnt in the past 10 years? *World J Gastroenterol* 20, 1192–210.

Hornef, M. (2015). Pathogens, commensal symbionts, and pathobionts: discovery and functional effects on the host. *ILAR J* 56, 159–62.

House, P.K., Vyas, A., and Sapolsky, R. (2011). Predator cat odors activate sexual arousal pathways in brains of *Toxoplasma gondii* infected rats. *PLoS One* 6, e23277.

Howard, D.L., Bush, G.L., and Breznak, J.A. (1985). The evolutionary significance of bacteria associated with *Rhagoletis*. *Evolution* 39, 405–17.

Huang, J.H., and Douglas, A.E. (2015). Consumption of dietary sugar by gut bacteria determines *Drosophila* lipid content. *Biol Lett* 11, 20150469.

Hudson, P.J., Dobson, A.P., and Lafferty, K.D. (2006). Is a healthy ecosystem one that is rich in parasites? *Trends Ecol Evol* 21, 381–85.

Huffnagle, G.B., and Noverr, M.C. (2013). The emerging world of the fungal microbiome. *Trends Microbiol* 21, 334–41.

Hulcr, J., and Stelinski, L.L. (2017). The ambrosia symbiosis: from evolutionary ecology to practical management. *Annu Rev Entomol* 62, 285–303.

Hullar, M.A., and Fu, B.C. (2014). Diet, the gut microbiome, and epigenetics. *Cancer J* 20, 170–75.

Human Microbiome Project Consortium (2012a). Structure, function and diversity of the healthy human microbiome. *Nature* 486, 207–14.

Human Microbiome Project Consortium (2012b). A framework for human microbiome research. *Nature* 486, 215–21.

Humann, J., Mann, B., Gao, G., Moresco, P., Ramahi, J., Loh, L.N., Farr, A., et al. (2016). Bacterial peptidoglycan traverses the placenta to induce fetal neuroproliferation and aberrant postnatal behavior. *Cell Host Microbe* 19, 901.

Hurst, J.L., Payne, C.E., Nevison, C.M., Marie, A.D., Humphries, R.E., Robertson, D.H., Cavaggioni, A., and Beynon, R.J. (2001). Individual recognition in mice mediated by major urinary proteins. *Nature* 414, 631–34.

Imeri, L., and Opp, M.R. (2009). How (and why) the immune system makes us sleep. *Nat Rev Neurosci* 10, 199–210.

Ivanov, I.I., Atarashi, K., Manel, N., Brodie, E.L., Shima, T., Karaoz, U., Wei, D., et al. (2009). Induction of intestinal Th17 cells by segmented filamentous bacteria. *Cell* 139, 485–98.

Jaenike, J., Unckless, R., Cockburn, S.N., Boelio, L.M., and Perlman, S.J. (2010). Adaptation via symbiosis: recent spread of a *Drosophila* defensive symbiont. *Science* 329, 212–15.

Jakobsson, H.E., Jernberg, C., Andersson, A.F., Sjolund-Karlsson, M., Jansson, J.K., and Engstrand, L. (2010). Short-term antibiotic treatment has differing long-term impacts on the human throat and gut microbiome. *PLoS One* 5, e9836.

Janeway, C.A. and Medzhitov, R. (2002). Innate immune recognition. *Ann Rev Immunol* 20, 197–216.

Jennings, D.H., ed. (1975). *Symbiosis* (Cambridge, UK: Cambridge University Press).

Johansson, M.E., Larsson, J.M., and Hansson, G.C. (2011). The two mucus layers of colon are organized by the MUC2 mucin, whereas the outer layer is a legislator of host-microbial interactions. *Proc Natl Acad Sci USA* 108 Suppl 1, 4659–65.

Joy, J.B. (2013). Symbiosis catalyses niche expansion and diversification. *Proc Biol Sci* 280, 20122820.

Kaasalainen, U., Fewer, D.P., Jokela, J., Wahlsten, M., Sivonen, K., and Rikkinen, J. (2012). Cyanobacteria produce a high variety of hepatotoxic peptides in lichen symbiosis. *Proc Natl Acad Sci USA* 109, 5886–91.

Kaltenpoth, M., Gottler, W., Herzner, G., and Strohm, E. (2005). Symbiotic bacteria protect wasp larvae from fungal infestation. *Curr Biol* 15, 475–79.

Kaltenpoth, M., Roeser-Mueller, K., Koehler, S., Peterson, A., Nechitaylo, T.Y., Stubblefield, J.W., Herzner, G., Seger, J., and Strohm, E. (2014). Partner choice and fidelity stabilize coevolution in a Cretaceous-age defensive symbiosis. *Proc Natl Acad Sci USA* 111, 6359–64

Kamada, N., Kim, Y.G., Sham, H.P., Vallance, B.A., Puente, J.L., Martens, E.C., Núñez, G. (2012). Regulated virulence controls the ability of a pathogen to compete with the gut microbiota. *Science* 336, 1325–29.

Karasov, W.H., and Douglas, A.E. (2013). Comparative digestive physiology. *Comp Physiol* 3, 741–83.

Kerney, R., Kim, E., Hangarter, R.P., Heiss, A.A., Bishop, C.D., and Hall, B.K. (2011). Intracellular invasion of green algae in a salamander host. *Proc Natl Acad Sci USA* 108, 6497–502.

Khosravi, A., Yáñez, A., Price, J.G., Chow, A., Merad, M., Goodridge, H.S., Mazmanian, S.K. (2014). Gut microbiota promote hematopoiesis to control bacterial infection. *Cell Host Microbe* 15, 374–81.

Kim, S.H., and Lee, W.J. (2014). Role of DUOX in gut inflammation: lessons from *Drosophila* model of gut-microbiota interactions. *Front Cell Infect Microbiol* 3, 16.

Kimura, I., Ozawa, K., Inoue, D., Imamura, T., Kimura, K., Maeda, T., Terasawa, K., et al. (2013). The gut microbiota suppresses insulin-mediated fat accumulation via the short-chain fatty acid receptor GPR43. *Nat Commun* 4, 1829.

Kliman, H.J. (2014). Comment on "The placenta harbors a unique microbiome." *Sci Transl Med* 6, 254le254.

Knights, D., Ward, T.L., McKinlay, C.E., Miller, H., Gonzalez, A., McDonald, D., and Knight, R. (2014). Rethinking "enterotypes." *Cell Host Microbe* 16, 433–37.

Knoll, A.H. (2014) Paleobiological perspectives on early eukaryotic evolution. *Cold Spring Harb Perspect Biol* 6, a016121.

Koeth, R.A., Wang, Z., Levison, B.S., Buffa, J.A., Org, E., Sheehy, B.T., Britt, E.B., et al. (2013). Intestinal microbiota metabolism of L-carnitine, a nutrient in red meat, promotes atherosclerosis. *Nat Med* 19, 576–85.

Kohl, K.D., and Dearing, M.D. (2016). The woodrat gut microbiota as an experimental system for understanding microbial metabolism of dietary toxins. *Front Microbiol* 7, 1165.

Kohl, K.D., Weiss, R.B., Cox, J., Dale, C., and Dearing, M.D. (2014). Gut microbes of mammalian herbivores facilitate intake of plant toxins. *Ecol Lett* 17, 1238–46.

Konopka, A. (2009). What is microbial community ecology? *ISME J* 3, 1223–30.

Koren, O., Spor, A., Felin, J., Fak, F., Stombaugh, J., Tremaroli, V., Behre, C.J., et al. (2011). Human oral, gut, and plaque microbiota in patients with atherosclerosis. *Proc Natl Acad Sci USA* 108 Suppl 1, 4592–98.

Koren, O., Goodrich, J.K., Cullender, T.C., Spor, A., Laitinen, K., Backhed, H.K., Gonzalez, A., et al. (2012). Host remodeling of the gut microbiome and metabolic changes during pregnancy. *Cell* 150, 470–80.

Koren, O., Knights, D., Gonzalez, A., Waldron, L., Segata, N., Knight, R., Huttenhower, C., and Ley, R.E. (2013). A guide to enterotypes across the human body: meta-analysis of microbial community structures in human microbiome datasets. *PLoS Comput Biol* 9, e1002863.

Koropatkin, N.M., Cameron, E.A., and Martens, E.C. (2012). How glycan metabolism shapes the human gut microbiota. *Nat Rev Microbiol* 10, 323–35.

Koropatnick T.A., Engle, J.T., Apicella, M.A., Stabb, E.V., Goldman, W.E., and McFall-Ngai, M.J. (2004). Microbial factor-mediated development in a host-bacterial mutualism. *Science* 306, 1186–88.

Korpela, K., Salonen, A., Virta, L.J., Kekkonen, R.A., Forslund, K., Bork, P., and de Vos, W.M. (2016). Intestinal microbiome is related to lifetime antibiotic use in Finnish pre-school children. *Nat Commun* 7, 10410.

Kostic, A.D., Chun, E., Robertson, L., Glickman, J.N., Gallini, C.A., Michaud, M., Clancy, T.E., et al. (2013). Fusobacterium nucleatum potentiates intestinal tumorigenesis and modulates the tumor-immune microenvironment. *Cell Host Microbe* 14, 207–15.

Kostic, A.D., Gevers, D., Siljander, H., Vatanen, T., Hyotylainen, T., Hamalainen, A.M., Peet, A., et al. (2015). The dynamics of the human infant gut microbiome in development and in progression toward type 1 diabetes. *Cell Host Microbe* 17, 260–73.

Koukou, K., Pavlikaki, H., Kilias, G., Werren, J.H., Bourtzis, K., and Alahiotis, S.N. (2006). Influence of antibiotic treatment and *Wolbachia* curing on sexual isolation among *Drosophila melanogaster* cage populations. *Evolution* 60, 87–96.

Kremer, N., Voronin, D., Charif, D., Mavingui, P., Mollereau, B., and Vavre, F. (2009). *Wolbachia* interferes with ferritin expression and iron metabolism in insects. *PLoS Pathog* 5, e1000630.

Kriegel, M.A., Sefik, E., Hill, J.A., Wu, H.J., Benoist, C., and Mathis, D. (2011). Naturally transmitted segmented filamentous bacteria segregate with diabetes protection in nonobese diabetic mice. *Proc Natl Acad Sci USA* 108, 11548–53.

Kroiss, J., Kaltenpoth, M., Schneider, B., Schwinger, M.G., Hertweck, C., Maddula, R.K., Strohm, E., and Svatos, A. (2010). Symbiotic streptomycetes provide antibiotic combination prophylaxis for wasp offspring. *Nature Chem Biol* 6, 261–63.

Krueger, J.M., Pappenheimer, J.R., and Karnovsky, M.L. (1982). The composition of sleep-promoting factor isolated from human urine. *J Biol Chem* 257, 1664–69.

Krych, L., Hansen, C.H., Hansen, A.K., van den Berg, F.W., and Nielsen, D.S. (2013). Quantitatively different, yet qualitatively alike: a meta-analysis of the mouse core gut microbiome with a view towards the human gut microbiome. *PLoS One* 8, e62578.

Kumar, H., Lund, R., Laiho, A., Lundelin, K., Ley, R.E., Isolauri, E., and Salminen, S. (2014). Gut microbiota as an epigenetic regulator: pilot study based on whole-genome methylation analysis. *mBio* 5, e02113–14.

Kunin, W.E., Vergeer, P., Kenta, T., Davey, M.P., Burke, T., Woodward, F.I., Quick, P., Mannarelli, M.E., Watson-Haigh, N.S., and Butlin, R. (2009). Variation at range margins across multiple spatial scales: environmental temperature, population genetics and metabolomic phenotype. *Proc Biol Sci* 276, 1495–506.

Kwan, J.C., Donia, M.S., Han, A.W., Hirose, E., Haygood, M.G., and Schmidt, E.W. (2012). Genome streamlining and chemical defense in a coral reef symbiosis. *Proc Natl Acad Sci USA* 109, 20655–60.

Lagkouvardos I., Pukall, R., Abt, B., Foesel, B.U., Meier-Kalthoff, J.P., Kumar, N., Bresciani, A., et al. (2016) The Mouse Intestinal Bacterial Collection (miBC) provides host-specific insight into cultured diversity and functional potential of the gut microbiota. *Nat Microbiol* 1, 16131

Langille, M.G., Zaneveld, J., Caporaso, J.G., McDonald, D., Knights, D., Reyes, J.A., Clemente, J.C., et al. (2013). Predictive functional profiling of microbial communities using 16S rRNA marker gene sequences. *Nat Biotechnol* 31, 814–21.

Lawrence, S.A., O'Toole, R., Taylor, M.W., and Davy, S.K. (2010). Subcuticular bacteria associated with two common New Zealand echinoderms: Characterization using 16S rRNA sequence analysis and fluorescence in situ hybridization. *Biol Bull* 218, 95–104.

Lederberg, J., and McCray, A. (2001). Ome sweet 'omics: a genealogical treasury of words. *The Scientist* 15, 8.

Lee, K.H., and Ruby, E.G. (1994). Effect of the squid host on the abundance and distribution of symbiotic *Vibrio fischeri* in nature. *Appl Environ Microbiol* 60, 1565–71.

Leonardo, T.E., and Mondor, E.B. (2006). Symbiont modifies host life-history traits that affect gene flow. *Proc Biol Sci* 273, 1079–84.

Leone, V., Gibbons, S.M., Martinez, K., Hutchison, A.L., Huang, E.Y., Cham, C.M., Pierre, J.F., et al. (2015). Effects of diurnal variation of gut microbes and high-fat feeding on host circadian clock function and metabolism. *Cell Host Microbe* 17, 681–89.

Ley, R.E., Lozupone, C.A., Hamady, M., Knight, R., and Gordon, J.I. (2008). Worlds within worlds: evolution of the vertebrate gut microbiota. *Nat Rev Microbiol* 6, 776–88.

Li, Q., Korzan, W.J., Ferrero, D.M., Chang, R.B., Roy, D.S., Buchi, M., Lemon, J.K., et al. (2013). Synchronous evolution of an odor biosynthesis pathway and behavioral response. *Curr Biol* 23, 11–20.

Lindquist, N., Barber, P.H., and Weisz, J.B. (2005). Episymbiotic microbes as food and defence for marine isopods: unique symbioses in a hostile environment. *Proc Biol Sci* 272, 1209–16

Login, F.H., Balmand, S., Vallier, A., Vincent-Monegat, C., Vigneron, A., Weiss-Gayet, M., Rochat, D., and Heddi, A. (2011). Antimicrobial peptides keep insect endosymbionts under control. *Science* 334, 362–65.

Lokmer, A., and Mathias Wegner, K. (2015). Hemolymph microbiome of Pacific oysters in response to temperature, temperature stress and infection. *ISME J* 9, 670–82.

Lowe, C.D., Minter, E.J., Cameron, D.D., and Brockhurst, M.A. (2016). Shining a Light on Exploitative Host Control in a Photosynthetic Endosymbiosis. *Curr Biol* 26, 207–11.

Luan, J.B., Chen, W., Hasegawa, D.K., Simmons, A.M., Wintermantel, W.M., Ling, K.S., Fei, Z., Liu, S.S., and Douglas, A.E. (2015). Metabolic coevolution in the bacterial symbiosis of whiteflies and related plant sap-feeding insects. *Genome Biol Evol* 7, 2635–47.

Luczynski, P., McVey Neufeld, K.A., Oriach, C.S., Clarke, G., Dinan, T.G., and Cryan, J.F. (2016). Growing up in a bubble: using germ-free animals to assess the influence of the gut microbiota on brain and behavior. *Int J Neuropsychopharmacol* 19, pyw020.

Lundberg, D.S., Lebeis, S.L., Paredes, S.H., Yourstone, S., Gehring, J., Malfatti, S., Tremblay, J., et al. (2012). Defining the core *Arabidopsis thaliana* root microbiome. *Nature* 488, 86–90.

Lysenko, E.S., Ratner, A.J., Nelson, A.L., Weiser, J.N. (2005). The role of innate immune responses in the outcome of interspecies competition for colonization of mucosal surfaces. *PLoS Pathog* 1, e1.

MacPhee, R.A., Hummelen, R., Bisanz, J.E., Miller, W.L., and Reid, G. (2010). Probiotic strategies for the treatment and prevention of bacterial vaginosis. *Expert Opin Pharmacother* 11, 2985–95.

Macpherson A.J., and Uhr T. (2004). Induction of protective IgA by intestinal dendritic cells carrying commensal bacteria. *Science* 303, 1662–65

Macpherson, A.J., and McCoy, K.D. (2015). Standardised animal models of host microbial mutualism. *Mucosal Immunol* 8, 476–86.

Mao, Y.K., Kasper, D.L., Wang, B., Forsythe, P., Bienenstock, J., and Kunze, W.A. (2013). *Bacteroides fragilis* polysaccharide A is necessary and sufficient for acute activation of intestinal sensory neurons. *Nat Commun* 4, 1465.

Margulis, L. (1991). Symbiogenesis and symbionticism. In *Symbiosis as a Source of Evolutionary Innovation*, L. Margulis, and R. Fester, eds. (Cambridge, Massachusetts: MIT Press), 1–14.

Mark Welch, J.L., Rossetti, B.J., Rieken, C.W., Dewhirst, F.E., and Borisy, G.G. (2016). Biogeography of a human oral microbiome at the micron scale. *Proc Natl Acad Sci USA* 113, E791–800.

Martin, W.F., Garg, S., and Zimorski, V. (2015). Endosymbiotic theories for eukaryote origin. *Philos Trans R Soc Lond B Biol Sci* 370, 20140330.

Martinez, J., Cogni, R., Cao, C., Smith, S., Illingworth, C.J., and Jiggins, F.M. (2016). Addicted? Reduced host resistance in populations with defensive symbionts. *Proc Biol Sci* 283, 20160778.

Martinez, J.L. (2009). Environmental pollution by antibiotics and by antibiotic resistance determinants. *Environ Pollut* 157, 2893–902.

Martinson, V.G., Danforth, B.N., Minckley, R.L., Rueppell, O., Tingek, S., and Moran, N.A. (2011). A simple and distinctive microbiota associated with honey bees and bumble bees. *Mol Ecol* 20, 619–28.

Matzinger, P. (2002). The danger model: a renewed sense of self. *Science* 296, 301–5.

Maynard Smith, J., and Harper, D. (2003). *Animal Signals* (Oxford, UK: Oxford University Press).

McCann, K.S. (2000). The diversity-stability debate. *Nature* 405, 228–33.

McCusker, R.H., and Kelley, K.W. (2013). Immune-neural connections: how the immune system's response to infectious agents influences behavior. *J Exp Biol* 216, 84–98.

McCutcheon, J.P., and von Dohlen, C.D. (2011). An interdependent metabolic patchwork in the nested symbiosis of mealybugs. *Curr Biol* 21, 1366–72.

McFadden, G.I. (2014). Origin and evolution of plastids and photosynthesis in eukaryotes. *Cold Spring Harb Perspect Biol* 6, a016105.

McFall-Ngai, M.J. (2002). Unseen forces: the influence of bacteria on animal development. *Dev Biol* 242, 1–14.

McFall-Ngai, M.J. (2007). Adaptive immunity: care for the community. *Nature* 445, 153.

McFall-Ngai, M., Hadfield, M.G., Bosch, T.C., Carey, H.V., Domazet-Loso, T., Douglas, A.E., Dubilier, N., et al. (2013). Animals in a bacterial world, a new imperative for the life sciences. *Proc Natl Acad Sci USA* 110, 3229–36.

McJunkin, B., Sissoko, M., Levien, J., Upchurch, J., and Ahmed, A. (2011). Dramatic decline in prevalence of *Helicobacter pylori* and peptic ulcer disease in an endoscopy-referral population. *Am J Med* 124, 260–64.

McLean, A.H., Parker, B.J., Hrcek, J., Henry, L.M., and Godfray, H.C. (2016). Insect symbionts in food webs. *Philos Trans R Soc Lond B Biol Sci* 371.

Meadow, J.F., Altrichter, A.E., Bateman, A.C., Stenson, J., Brown, G.Z., Green, J.L., and Bohannan, B.J. (2015). Humans differ in their personal microbial cloud. *PeerJ* 3, e1258.

Mekkes, M.C., Weenen, T.C., Brummer, R.J., and Claassen, E. (2014). The development of probiotic treatment in obesity: a review. *Benef Microbes* 5, 19–28.

Messaoudi, M., Lalonde, R., Violle, N., Javelot, H., Desor, D., Nejdi, A., Bisson, J.F., et al. (2011). Assessment of psychotropic-like properties of a probiotic formulation (*Lactobacillus helveticus* R0052 and *Bifidobacterium longum* R0175) in rats and human subjects. *Br J Nutr* 105, 755–64.

Metcalf, J.L., Xu, Z.Z., Weiss, S., Lax, S., Van Treuren, W., Hyde, E.R., Song, S.J., et al. (2016). Microbial community assembly and metabolic function during mammalian corpse decomposition. *Science* 351, 158–62

Metwalli, K.H., Khan, S.A., Krom, B.P., and Jabra-Rizk, M.A. (2013). *Streptococcus mutans*, *Candida albicans*, and the human mouth: a sticky situation. *PLoS Pathog* 9, e1003616.

Mikaelyan, A., Thompson, C.L., Hofer, M.J., and Brune, A. (2016). The deterministic assembly of complex bacterial communities in germ-free cockroach guts. *Appl Environ Microbiol* 82, 1256–63.

Miller, C.P., Bonhoff, M., and Rifkind, D. (1957). The effect of an antibiotic on the susceptibility of the mouse's intestinal tract to *Salmonella* infection. *Trans Am Clin Climatol Assoc* 68, 51–58

Modi, S.R., Lee, H.H., Spina, C.S., and Collins, J.J. (2013). Antibiotic treatment expands the resistance reservoir and ecological network of the phage metagenome. *Nature* 499, 219–22.

Moeller, A.H., Caro-Quintero, A., Mjungu, D., Georgiev, A.V., Lonsdorf, E.V., Muller, M.N., Pusey, A.E., Peeters, M., Hahn, B.H., and Ochman, H. (2016). Cospeciation of gut microbiota with hominids. *Science* 353, 380–82.

Molmeret, M., Horn, M., Wagner, M., Santic, M., and Abu Kwaik, Y. (2005). Amoebae as training grounds for intracellular bacterial pathogens. *Appl Environ Microbiol* 71, 20–28.

Moran, N.A. (1996). Accelerated evolution and Muller's ratchet in endosymbiotic bacteria. *Proc Natl Acad Sci USA* 93, 2873–78.

Moran, N.A., and Sloan, D.B. (2015). The hologenome concept: helpful or hollow? *PLoS Biol* 13, e1002311.

Moritz, C., and Agudo, R. (2013). The future of species under climate change: resilience or decline? *Science* 341, 504–8.

Morris, B.E., Henneberger, R., Huber, H., and Moissl-Eichinger, C. (2013). Microbial syntrophy: interaction for the common good. *FEMS Microbiol Rev* 37, 384–406.

Morris, J.J., Lenski, R.E., Zinser, E.R. (2012). The Black Queen Hypothesis: evolution of dependencies through adaptive gene loss. *mBio* 3, e00036–12.

Mueller, N.T., Bakacs, E., Combellick, J., Grigoryan, Z., and Dominguez-Bello, M.G. (2015). The infant microbiome development: mom matters. *Trends Mol Med* 21, 109–17.

Nakayama, T., Kamikawa, R., Tanifuji, G., Kashiyama, Y., Ohkouchi, N., Archibald, J.M., and Inagaki, Y. (2014). Complete genome of a nonphotosynthetic cyanobacterium in a diatom reveals recent adaptations to an intracellular lifestyle. *Proc Natl Acad Sci USA* 111, 11407–12.

Narayanan, N.S., Guarnieri, D.J., and DiLeone, R.J. (2010). Metabolic hormones, dopamine circuits, and feeding. *Front Neuroendocrinol* 31, 104–12.

Neu, J., and Rushing, J. (2011). Cesarean versus vaginal delivery: long-term infant outcomes and the hygiene hypothesis. *Clin Perinatol* 38, 321–31.

Nicholson, J.K., Holmes, E., Kinross, J., Burcelin, R., Gibson, G., Jia, W., and Pettersson, S. (2012). Host-gut microbiota metabolic interactions. *Science* 336, 1262–67.

Nieuwdorp, M., Gilijamse, P.W., Pai, N., and Kaplan, L.M. (2014). Role of the microbiome in energy regulation and metabolism. *Gastroenterol* 146, 1525–33.

Nikoh, N., McCutcheon, J.P., Kudo, T., Miyagishima, S.Y., Moran, N.A., and Nakabachi, A. (2010). Bacterial genes in the aphid genome: absence of functional gene transfer from *Buchnera* to its host. *PLoS Genet* 6, e1000827.

Nikoh, N., Tanaka, K., Shibata, F., Kondo, N., Hizume, M., Shimada, M., and Fukatsu, T. (2008). *Wolbachia* genome integrated in an insect chromosome: evolution and fate of laterally transferred endosymbiont genes. *Genome Res* 18, 272–80.

Nguyen, T.L., Vieira-Silva, S., Liston, A., and Raes, J. (2015). How informative is the mouse for human gut microbiota research? *Dis Model Mech* 8, 1–16.

Noble, D. (2013). Physiology is rocking the foundations of evolutionary biology. *Exp Physiol* 98, 1235–43.

Noble, R., Dobrovin-Pennington, A., Hobbs, P.J., Pederby, J., and Rodger, A. (2009). Volatile C8 compounds and pseudomonads influence primordium formation of *Agaricus bisporus*. *Mycologia* 101, 583–91.

Nowack, E.C., and Melkonian, M. (2010). Endosymbiotic associations within protists. *Philos Trans R Soc Lond B Biol Sci* 365, 699–712.

Nowack, E.C., Price, D.C., Bhattacharya, D., Singer, A., Melkonian, M. and Grossman, A.R. (2016). Gene transfers from diverse bacteria compensate for reductive genome evolution in the chromatophore of *Paulinella chromatophora*. *Proc Natl Acad Sci USA* 113, 12214–19.

Nowak, M.A. (2006). Five rules for the evolution of cooperation. *Science* 314, 1560–63.

Nutman, P.S., and Mosse, B., eds. (1963). *Symbiotic Associations* (Cambridge, UK: Cambridge University Press).

Nutzmann, H.W., Reyes-Dominguez, Y., Scherlach, K., Schroeckh, V., Horn, F., Gacek, A., Schumann, J., Hertweck, C., Strauss, J., and Brakhage, A.A. (2011). Bacteria-induced natural product formation in the fungus *Aspergillus nidulans* requires Saga/Ada-mediated histone acetylation. *Proc Natl Acad Sci USA* 108, 14282–87.

Nyholm, S.V., Stewart, J.J., Ruby, E.G., and McFall-Ngai, M.J. (2009). Recognition between symbiotic *Vibrio fischeri* and the haemocytes of *Euprymna scolopes*. *Environ Microbiol* 11, 483–93.

Obeng, N., Pratama, A.A., and Elsas, J.D. (2016). The significance of mutualistic phages for bacterial ecology and evolution. *Trends Microbiol* 24, 440–49.

Ochman, H., Worobey, M., Kuo, C.H., Ndjango, J.B., Peeters, M., Hahn, B.H., and Hugenholtz, P. (2010). Evolutionary relationships of wild hominids recapitulated by gut microbial communities. *PLoS Biol* 8, e1000546.

Oh, J., Byrd, A.L., Deming, C., Conlan, S., Program, N.C.S., Kong, H.H., and Segre, J.A. (2014). Biogeography and individuality shape function in the human skin metagenome. *Nature* 514, 59–64.

Oh, J., Byrd, A.L., Park, M., Program, N.C.S., Kong, H.H., and Segre, J.A. (2016). Temporal stability of the human skin microbiome. *Cell* 165, 854–66.

Oh, P.L., Benson, A.K., Peterson, D.A., Patil, P.B., Moriyama, E.N., Roos, S., and Walter, J. (2010). Diversification of the gut symbiont *Lactobacillus reuteri* as a result of host-driven evolution. *ISME J* 4, 377–87.

Oliveira, J.H., Gonçalves, R.L., Lara, F.A., Dias, F.A., Gandara, A.C., Menna-Barreto, R.F., Edwards, M.C., et al. (2011). Blood meal-derived heme decreases ROS levels in the midgut of *Aedes aegypti* and allows proliferation of intestinal microbiota. *PLoS Pathog* 7, e1001320

Oliveira, N.M., Niehus, R., and Foster, K.R. (2014). Evolutionary limits to cooperation in microbial communities. *Proc Natl Acad Sci USA* 111, 17941–46.

Oliver, K.M., Campos, J., Moran, N.A., and Hunter, M.S. (2008). Population dynamics of defensive symbionts in aphids. *Proc Biol Sci* 275, 293–99.

Oliver, K.M., Degnan, P.H., Burke, G.R., and Moran, N.A. (2010). Facultative symbionts in aphids and the horizontal transfer of ecologically important traits. *Annu Rev Entomol* 55, 247–66.

Olszak, T., An, D., Zeissig, S., Vera, M.P., Richter, J., Franke, A., Glickman, J.N., et al. (2012). Microbial exposure during early life has persistent effects on natural killer T cell function. *Science* 336, 489–93

O'Malley, M.A. (2015). Endosymbiosis and its implications for evolutionary theory. *Proc Natl Acad Sci USA* 112, 10270–10277.

Ost, A., Lempradl, A., Casas, E., Weigert, M., Tiko, T., Deniz, M., Pantano, L., et al. (2014). Paternal diet defines offspring chromatin state and intergenerational obesity. *Cell* 159, 1352–64.

Overend, G., Luo, Y., Henderson, L., Douglas, A.E., Davies, S.A., and Dow, J.A. (2016). Molecular mechanism and functional significance of acid generation in the *Drosophila* midgut. *Sci Rep* 6, 27242.

Padje, A.V., Whiteside, M.D., and Kiers, E.T. (2016). Signals and cues in the evolution of plant-microbe communication. *Curr Opin Plant Biol* 32, 47–52.

Palmer, C., Bik, E.M., DiGiulio, D.B., Relman, D.A., and Brown, P.O. (2007). Development of the human infant intestinal microbiota. *PLoS Biol* 5, e177.

Pande, S., Shitut, S., Freund, L., Westermann, M., Bertels, F., Colesie, C., Bischofs, I.B., and Kost, C. (2015). Metabolic cross-feeding via intercellular nanotubes among bacteria. *Nat Commun* 6, 6238.

Pannebakker, B.A., Loppin, B., Elemans, C.P., Humblot, L., and Vavre, F. (2007). Parasitic inhibition of cell death facilitates symbiosis. *Proc Natl Acad Sci USA* 104, 213–15.

Parsek, M.R., and Greenberg, E.P. (2005). Sociomicrobiology: the connections between quorum sensing and biofilms. *Trends Microbiol* 13, 27–33.

Pasteur, L. (1885). Observation relative a la note precedente de M. Duciaux. *Comptes Rendus de l'Académie des Sciences* 100, 68.

Perez-Burgos, A., Wang, B., Mao, Y.K., Mistry, B., McVey Neufeld, K.A., Bienenstock, J., and Kunze, W. (2013). Psychoactive bacteria *Lactobacillus rhamnosus* (JB-1) elicits rapid frequency facilitation in vagal afferents. *Am J Physiol Gastrointest Liver Physiol* 304, G211–20.

Perez-Cobas, A.E., Gosalbes, M.J., Friedrichs, A., Knecht, H., Artacho, A., Eismann, K., Otto, W., et al. (2013). Gut microbiota disturbance during antibiotic therapy: a multi-omic approach. *Gut* 62, 1591–601.

Perlman, S.J., Hodson, C.N., Hamilton, P.T., Opit, G.P., and Gowen, B.E. (2015). Maternal transmission, sex ratio distortion, and mitochondria. *Proc Natl Acad Sci USA* 112, 10162–68.

Perry, B., and Wang, Y. (2012). Appetite regulation and weight control: the role of gut hormones. *Nutr Diabetes* 2, e26.

Perry, R.J., Peng, L., Barry, N.A., Cline, G.W., Zhang, D., Cardone, R.L., Petersen, K.F., Kibbey, R.G., Goodman, A.L., and Shulman, G.I. (2016). Acetate mediates a microbiome-brain-beta-cell axis to promote metabolic syndrome. *Nature* 534, 213–17.

Peterson, G., Kumar, A., Gart, E., and Narayanan, S. (2011). Catecholamines increase conjugative gene transfer between enteric bacteria. *Microb Pathog* 51, 1–8.

Petroni, G., Spring, S., Schleifer, K.H., Verni, F., and Rosati, G. (2000). Defensive extrusive ectosymbionts of *Euplotidium* (Ciliophora) that contain microtubule-like structures are bacteria related to Verrucomicrobia. *Proc Natl Acad Sci USA* 97, 1813–17.

Pickard, J.M., Maurice, C.F., Kinnebrew, M.A., Abt, M.C., Schenten, D., Golovkina, T.V., Bogatyrev, S.R., et al. (2014). Rapid fucosylation of intestinal epithelium sustains host-commensal symbiosis in sickness. *Nature* 514, 638–41

Piel, J. (2002). A polyketide synthase-peptide synthetase gene cluster from an uncultured bacterial symbiont of *Paederus* beetles. *Proc Natl Acad Sci USA* 99, 14002–7

Pool, A.H., and Scott, K. (2014). Feeding regulation in *Drosophila*. *Curr Opin Neurobiol* 29, 57–63.

Qin, J., Li, R., Raes, J., Arumugam, M., Burgdorf, K.S., Manichanh, C., Nielsen, T., et al. (2010). A human gut microbial gene catalogue established by metagenomic sequencing. *Nature* 464, 59–65.

Qin, J., Li, Y., Cai, Z., Li, S., Zhu, J., Zhang, F., Liang, S., et al. (2012). A metagenome-wide association study of gut microbiota in type 2 diabetes. *Nature* 490, 55–60.

Rainey, P.B., and Rainey, K. (2003). Evolution of cooperation and conflict in experimental bacterial populations. *Nature* 425, 72–74.

Rajpal, D.K., and Brown, J.R. (2013). Modulating the human gut microbiome as an emerging therapeutic paradigm. *Sci Prog* 96, 224–36.

Rajilic-Stojanovic, M., Biagi, E., Heilig, H.G., Kajander, K., Kekkonen, R.A., Tims, S., and de Vos, W.M. (2011). Global and deep molecular analysis of microbiota signatures in fecal samples from patients with irritable bowel syndrome. *Gastroenterol* 141, 1792–801.

Rakoff-Nahoum, S., Foster, K.R., and Comstock, L.E. (2016). The evolution of cooperation within the gut microbiota. *Nature* 533, 255–59.

Ravel, J., Gajer, P., Abdo, Z., Schneider, G.M., Koenig, S.S., McCulle, S.L., Karlebach, S., et al. (2011). Vaginal microbiome of reproductive-age women. *Proc Natl Acad Sci USA* 108 Suppl 1, 4680–87.

Reuter, M., Bell, G., and Greig, D. (2007). Increased outbreeding in yeast in response to dispersal by an insect vector. *Curr Biol* 17, R81–83.

Reyes, A., Wu, M., McNulty, N.P., Rohwer, F.L., and Gordon, J.I. (2013). Gnotobiotic mouse model of phage-bacterial host dynamics in the human gut. *Proc Natl Acad Sci USA* 110, 20236–41.

Ribeiro, F.J., Przybylski, D., Yin, S., Sharpe, T., Gnerre, S., Abouelleil, A., Berlin, A.M., et al. (2012). Finished bacterial genomes from shotgun sequence data. *Genome Res* 22, 2270–77.

Ribes, M., Jimenez, E., Yahel, G., Lopez-Sendino, P., Diez, B., Massana, R., Sharp, J.H., and Coma, R. (2012). Functional convergence of microbes associated with temperate marine sponges. *Environ Microbiol* 14, 1224–39.

Ridaura, V.K., Faith, J.J., Rey, F.E., Cheng, J., Duncan, A.E., Kau, A.L., Griffin, N.W., et al. (2013). Gut microbiota from twins discordant for obesity modulate metabolism in mice. *Science* 341, 1241214.

Riedel, K., Hentzer, M., Geisenberger, O., Huber, B., Steidle, A., Wu, H., Hoiby, N., Givskov, M., Molin, S., and Eberl, L. (2001). N-acylhomoserine-lactone-mediated communication between *Pseudomonas aeruginosa* and *Burkholderia cepacia* in mixed biofilms. *Microbiol* 147, 3249–62.

Ritchie, K.B. (2006). Regulation of microbial populations by coral surface mucus and mucus-associated bacteria. *Mar Ecol Prog Series* 322, 1–14.

Robson, S.J., Vally, H., Abdel-Latif, M.E., Yu, M., and Westrupp, E. (2015). Childhood health and developmental outcomes after Cesarean birth in an Australian cohort. *Pediatrics* 136, e1285–93.

Rodriguez, J.M., Murphy, K., Stanton, C., Ross, R.P., Kober, O.I., Juge, N., Avershina, E., et al. (2015). The composition of the gut microbiota throughout life, with an emphasis on early life. *Microb Ecol Health Dis* 26, 26050.

Rohde, C.M., Wells, D.F., Robosky, L.C., Manning, M.L., Clifford, C.B., Reily, M.D., and Robertson, D.G. (2007). Metabonomic evaluation of Schaedler altered microflora rats. *Chem Res Toxicol* 20, 1388–92.

Rolff, J., and Schmid-Hempel, P. (2016). Perspectives on the evolutionary ecology of arthropod antimicrobial peptides. *Philos Trans R Soc Lond B Biol Sci* 371, 20150292.

Rolig, A.S., Parthasarathy, R., Burns, A.R., Bohannan, B.J., and Guillemin, K. (2015). Individual members of the microbiota disproportionately modulate host innate immune responses. *Cell Host Microbe* 18, 613–20.

Rook, G.A., and Brunet, L.R. (2005). Microbes, immunoregulation, and the gut. *Gut* 54, 317–20.

Rosenberg, E., and Zilber-Rosenberg, I. (2016). Microbes drive evolution of animals and plants: the hologenome concept. *mBio* 7, e01395.

Round, J.L., and Mazmanian, S.K. (2009). The gut microbiota shapes intestinal immune responses during health and disease. *Nat Rev Immunol* 9, 313–23.

Round, J.L., and Mazmanian, S.K. (2010). Inducible Foxp3+ regulatory T-cell development by a commensal bacterium of the intestinal microbiota. *Proc Natl Acad Sci USA* 107, 12204–9.

Rouphael, N.G., and Stephens, D.S. (2012). *Neisseria meningitidis*: biology, microbiology, and epidemiology. *Methods Mol Biol* 799, 1–20.

Rowan, R., Knowlton, N., Baker, A., and Jara, J. (1997). Landscape ecology of algal symbionts creates variation in episodes of coral bleaching. *Nature* 388, 265–69.

Russell, C.W., Bouvaine, S., Newell, P.D., and Douglas, A.E. (2013). Shared metabolic pathways in a coevolved insect-bacterial symbiosis. *Appl Environ Microbiol* 79, 6117–23.

Russell, C.W., Poliakov, A., Haribal, M., Jander, G., van Wijk, K.J., and Douglas, A.E. (2014). Matching the supply of bacterial nutrients to the nutritional demand of the animal host. *Proc Biol Sci* 281, 20141163.

Ryu, J.H., Kim, S.H., Lee, H.Y., Bai, J.Y., Nam, Y.D., Bae, J.W., Lee, D.G., Shin, S.C., Ha, E.M., and Lee, W.J. (2008). Innate immune homeostasis by the homeobox gene caudal and commensal-gut mutualism in *Drosophila*. *Science* 319, 777–82.

Sadd, B.M., and Schmid-Hempel, P. (2009). Principles of ecological immunology. *Evol Appl* 2, 113–21.

Salmela, H., Amdam, G.V., and Freitak, D. (2015). Transfer of immunity from mother to offspring is mediated via egg-yolk protein vitellogenin. *PLoS Pathog* 11, e1005015.

Sang, J.H. (1956). The quantitative nutritional requirements of *Drosophila melanogaster. J Exp Biol* 33, 45–72.

Sansone, C.L., Cohen, J., Yasunaga, A., Xu, J., Osborn, G., Subramanian, H., Gold, B., Buchon, N., and Cherry, S. (2015). Microbiota-dependent priming of antiviral intestinal immunity in *Drosophila. Cell Host Microbe* 18, 571–81.

Sapp, J. (1994). *Evolution by Association* (Oxford, UK: Oxford University Press).

Scher, J.U., Sczesnak, A., Longman, R.S., Segata, N., Ubeda, C., Bielski, C., Rostron, T., et al. (2013). Expansion of intestinal *Prevotella copri* correlates with enhanced susceptibility to arthritis. *Elife* 2, e01202.

Scherlach, K., Busch, B., Lackner, G., Paszkowski, U., and Hertweck, C. (2012). Symbiotic cooperation in the biosynthesis of a phytotoxin. *Angew Chem Int Ed Engl* 51, 9615–18.

Schirbel, A., Kessler, S., Rieder, F., West, G., Rebert, N., Asosingh, K., McDonald, C., and Fiocchi, C. (2013). Pro-angiogenic activity of TLRs and NLRs: a novel link between gut microbiota and intestinal angiogenesis. *Gastroenterol* 144, 613–23 e619.

Schloss, P.D., Iverson, K.D., Petrosino, J.F., and Schloss, S.J. (2014). The dynamics of a family's gut microbiota reveal variations on a theme. *Microbiome* 2, 25.

Schmidt, O., Pfanner, N., and Meisinger, C. (2010). Mitochondrial protein import: from proteomics to functional mechanisms. *Nat Rev Mol Cell Biol* 11, 655–67.

Schmitz, A., Anselme, C., Ravallec, M., Rebuf, C., Simon, J.C., Gatti, J.L., and Poirié, M. (2012). The cellular immune response of the pea aphid to foreign intrusion and symbiotic challenge. *PLoS One* 7, e42114.

Schnorr, S.L., Candela, M., Rampelli, S., Centanni, M., Consolandi, C., Basaglia, G., Turroni, S., et al. (2014). Gut microbiome of the Hadza hunter-gatherers. *Nat Commun* 5, 3654.

Schulz, F., and Horn, M. (2015). Intranuclear bacteria: inside the cellular control center of eukaryotes. *Trends Cell Biol* 25, 339–46.

Schwab, U., Abdullah, L.H., Perlmutt, O.S., Albert, D., Davis, C.W., Arnold, R.R., Yankaskas, J.R., et al. (2014). Localization of *Burkholderia cepacia* complex bacteria in cystic fibrosis lungs and interactions with *Pseudomonas aeruginosa* in hypoxic mucus. *Infect Immun* 82, 4729–45.

Schwabe, R.F., and Jobin, C. (2013). The microbiome and cancer. *Nat Rev Cancer* 13, 800–812.

Schwarcz, R., Bruno, J.P., Muchowski, P.J., and Wu, H.Q. (2012). Kynurenines in the mammalian brain: when physiology meets pathology. *Nat Rev Neurosci* 13, 465–77.

Sears, C.L., and Garrett, W.S. (2014). Microbes, microbiota, and colon cancer. *Cell Host Microbe* 15, 317–28.

Sela, D.A., and Mills, D.A. (2010). Nursing our microbiota: molecular linkages between bifidobacteria and milk oligosaccharides. *Trends Microbiol* 18, 298–307.

Sender, R., Fuchs, S., and Milo, R. (2016). Revised estimates for the number of human and bacteria cells in the body. *PLoS Biol* 14, e1002533.

Shade, A., and Handelsman, J. (2012). Beyond the Venn diagram: the hunt for a core microbiome. *Environ Microbiol* 14, 4–12.

Sharon, G., Segal, D., Ringo, J.M., Hefetz, A., Zilber-Rosenberg, I., and Rosenberg, E. (2010). Commensal bacteria play a role in mating preference of *Drosophila melanogaster. Proc Natl Acad Sci USA* 107, 20051–56.

Shin, S.C., Kim, S.H., You, H., Kim, B., Kim, A.C., Lee, K.A., Yoon, J.H., Ryu, J.H., and Lee, W.J. (2011). *Drosophila* microbiome modulates host developmental and metabolic homeostasis via insulin signaling. *Science* 334, 670–74.

Shropshire, J.D., and Bordenstein, S.R. (2016). Speciation by symbiosis: the microbiome and behavior. *mBio* 7, e01785.

Sloan, D.B., Nakabachi, A., Richards, S., Qu, J., Murali, S.C., Gibbs, R.A., and Moran, N.A. (2014). Parallel histories of horizontal gene transfer facilitated extreme reduction of endosymbiont genomes in sap-feeding insects. *Mol Biol Evol* 31, 857–71.

Smillie, C.S., Smith, M.B., Friedman, J., Cordero, O.X., David, L.A., and Alm, E.J. (2011). Ecology drives a global network of gene exchange connecting the human microbiome. *Nature* 480, 241–44.

Smilowitz, J.T., Lebrilla, C.B., Mills, D.A., German, J.B., and Freeman, S.L. (2014). Breast milk oligosaccharides: structure-function relationships in the neonate. *Annu Rev Nutr* 34, 143–69.

Smith, C.R., Glover, A.G., Treude, T., Higgs, N.D., and Amon, D.J. (2015). Whale-fall ecosystems: recent insights into ecology, paleoecology, and evolution. *Ann Rev Mar Sci* 7, 571–96.

Smith, K., McCoy, K.D., and Macpherson, A.J. (2007). Use of axenic animals in studying the adaptation of mammals to their commensal intestinal microbiota. *Sem Immunol* 19, 59–69.

Smith, M.I., Yatsunenko, T., Manary, M.J., Trehan, I., Mkakosya, R., Cheng, J., Kau, A.L., et al. (2013). Gut microbiomes of Malawian twin pairs discordant for kwashiorkor. *Science* 339, 548–54.

Smith, P.M., Howitt, M.R., Panikov, N., Michaud, M., Gallini, C.A., Bohlooly-Y, M., Glickman, J.N., and Garrett, W.S. (2013). The microbial metabolites, short-chain fatty acids, regulate colonic Treg cell homeostasis. *Science* 341, 569–73.

Sohn, J.W., Elmquist, J.K., and Williams, K.W. (2013). Neuronal circuits that regulate feeding behavior and metabolism. *Trends Neurosci* 36, 504–12.

Sokol, H., Pigneur, B., Watterlot, L., Lakhdari, O., Bermudez-Humaran, L.G., Gratadoux, J.J., Blugeon, S., et al. (2008). *Faecalibacterium prausnitzii* is an anti-inflammatory commensal bacterium identified by gut microbiota analysis of Crohn disease patients. *Proc Natl Acad Sci USA* 105, 16731–36.

Sommer, F., Stahlman, M., Ilkayeva, O., Arnemo, J.M., Kindberg, J., Josefsson, J., Newgard, C.B., Frobert, O., and Backhed, F. (2016). The gut microbiota modulates energy metabolism in the hibernating brown bear *Ursus arctos*. *Cell Rep* 14, 1655–61.

Song, S.J., Lauber, C., Costello, E.K., Lozupone, C.A., Humphrey, G., Berg-Lyons, D., Caporaso, J.G., et al. (2013). Cohabiting family members share microbiota with one another and with their dogs. *Elife* 2, e00458.

Sonnenburg, E.D., Smits, S.A., Tikhonov, M., Higginbottom, S.K., Wingreen, N.S., and Sonnenburg, J.L. (2016). Diet-induced extinctions in the gut microbiota compound over generations. *Nature* 529, 212–15.

Spang, A., Saw, J.H., Jorgensen, S.L., Zaremba-Niedzwiedzka, K., Martijn, J., Lind, A.E., van Eijk, R., Schleper, C., Guy, L., and Ettema, T.J. (2015). Complex archaea that bridge the gap between prokaryotes and eukaryotes. *Nature* 521, 173–79.

Spribille, T., Tuovinen, V., Resl, P., Vanderpool, D., Wolinski, H., Aime, M.C., Schneider, K., et al. (2016). Basidiomycete yeasts in the cortex of ascomycete macrolichens. *Science* 353, 488–92.

Srivastava, M., Simakov, O., Chapman, J., Fahey, B., Gauthier, M.E., Mitros, T., Richards, G.S., et al. (2010). The *Amphimedon queenslandica* genome and the evolution of animal complexity. *Nature* 466, 720–26.

Stappenbeck, T.S., Hooper, L.V., and Gordon, J.I. (2002). Developmental regulation of intestinal angiogenesis by indigenous microbes via Paneth cells. *Proc Natl Acad Sci USA* 99, 15451–55.

Stecher, B., Denzler, R., Maier, L., Bernet, F., Sanders, M.J., Pickard, D.J., Barthel, M., et al. (2012). Gut inflammation can boost horizontal gene transfer between pathogenic and commensal Enterobacteriaceae. *Proc Natl Acad Sci USA* 109, 1269–74.

Stecher, B., Maier, L., and Hardt, W.D. (2013). 'Blooming' in the gut: how dysbiosis might contribute to pathogen evolution. *Nat Rev Microbiol* 11, 277–84.

Stecher, B., Robbiani, R., Walker, A.W., Westendorf, A.M., Barthel, M., Kremer, M., Chaffron, S., et al. (2007). *Salmonella enterica* serovar typhimurium exploits inflammation to compete with the intestinal microbiota. *PLoS Biol* 5, 2177–89

Stefanini, I., Dapporto, L., Berna, L., Polsinelli, M., Turillazzi, S., and Cavalieri, D. (2016). Social wasps are a *Saccharomyces* mating nest. *Proc Natl Acad Sci USA* 113, 2247–51.

Stefanini, I., Dapporto, L., Legras, J.L., Calabretta, A., Di Paola, M., De Filippo, C., Viola, R., et al. (2012). Role of social wasps in *Saccharomyces cerevisiae* ecology and evolution. *Proc Natl Acad Sci USA* 109, 13398–403.

Stein, R.R., Bucci, V., Toussaint, N.C., Buffie, C.G., Ratsch, G., Pamer, E.G., Sander, C., and Xavier, J.B. (2013). Ecological modeling from time-series inference: insight into dynamics and stability of intestinal microbiota. *PLoS Comput Biol* 9, e1003388.

Strachan, D.P. (1989). Hay fever, hygiene, and household size. *BMJ* 299, 1259–60.

Sudo, N., Chida, Y., Aiba, Y., Sonoda, J., Oyama, N., Yu, X.N., Kubo, C., and Koga, Y. (2004). Postnatal microbial colonization programs the hypothalamic-pituitary-adrenal system for stress response in mice. *J Physiol* 558, 263–75.

Suzuki, K., Meek, B., Doi, Y., Muramatsu, M., Chiba, T., Honjo, T., and Fagarasan, S. (2004). Aberrant expansion of segmented filamentous bacteria in IgA-deficient gut. *Proc Natl Acad Sci USA* 101, 1981–86.

Taborsky, M. (2013). Social evolution: reciprocity there is. *Curr Biol* 23, R486–88.

Takahashi, S., Kapas, L., and Krueger, J.M. (1996). A tumor necrosis factor (TNF) receptor fragment attenuates TNF-alpha- and muramyl dipeptide-induced sleep and fever in rabbits. *J Sleep Res* 5, 106–14.

Teixeira, L., Ferreira, A., and Ashburner, M. (2008). The bacterial symbiont *Wolbachia* induces resistance to RNA viral infections in *Drosophila melanogaster*. *PLoS Biol* 6, e2.

Telfer, S., Bown, K.J., Sekules, R., Begon, M., Hayden, T., and Birtles, R. (2005). Disruption of a host-parasite system following the introduction of an exotic host species. *Parasitol* 130, 661–68.

Theis, K.R., Schmidt, T.M., and Holekamp, K.E. (2012). Evidence for a bacterial mechanism for group-specific social odors among hyenas. *Sci Rep* 2, 615.

Theis, K.R., Venkataraman, A., Dycus, J.A., Koonter, K.D., Schmitt-Matzen, E.N., Wagner, A.P., Holekamp, K.E., and Schmidt, T.M. (2013). Symbiotic bacteria appear to mediate hyena social odors. *Proc Natl Acad Sci USA* 110, 19832–37.

Thimm, T., Hoffmann, A., Fritz, I., and Tebbe, C.C. (2001). Contribution of the earthworm *Lumbricus rubellus* (Annelida, Oligochaeta) to the establishment of plasmids in soil bacterial communities. *Microb Ecol* 41, 341–51.

Thomas, G.H., Zucker, J., Macdonald, S.J., Sorokin, A., Goryanin, I., and Douglas, A.E. (2009). A fragile metabolic network adapted for cooperation in the symbiotic bacterium *Buchnera aphidicola*. *BMC Syst Biol* 3, 24.

Thomas, T., Moitinho-Silva, L., Lurgi, M., Bjork, J.R., Easson, C., Astudillo-Garcia, C., Olson, J.B., et al. (2016). Diversity, structure and convergent evolution of the global sponge microbiome. *Nat Commun* 7, 11870.

Thubaut, J., Puillandre, N., Faure, B., Cruaud, C., and Samadi, S. (2013). The contrasted evolutionary fates of deep-sea chemosynthetic mussels (Bivalvia, Bathymodiolinae). *Ecol Evol* 3, 4748–66.

Tillisch, K., Labus, J., Kilpatrick, L., Jiang, Z., Stains, J., Ebrat, B., Guyonnet, D., et al. (2013). Consumption of fermented milk product with probiotic modulates brain activity. *Gastroenterol* 144, 1394–401, 1401.e1–4.

Tremblay, P., Grover, R., Maguer, J.F., Legendre, L., and Ferrier-Pages, C. (2012). Autotrophic carbon budget in coral tissue: a new 13C-based model of photosynthate translocation. *J Exp Biol* 215, 1384–93.

Tsavkelova, E.A., Klimova, S., Cherdyntseva, T.A., and Netrusov, A.I. (2006). Hormones and hormone-like substances of microorganisms: a review. *Prikl Biokhim Mikrobiol* 42, 261–68.

Tung, J., Barreiro, L.B., Burns, M.B., Grenier, J.C., Lynch, J., Grieneisen, L.E., Altmann, J., Alberts, S.C., Blekhman, R., and Archie, E.A. (2015). Social networks predict gut microbiome composition in wild baboons. *Elife* 4, e05224.

Turnbaugh, P.J., Ley, R.E., Mahowald, M.A., Magrini, V., Mardis, E.R., and Gordon, J.I. (2006). An obesity-associated gut microbiome with increased capacity for energy harvest. *Nature* 444, 1027–31.

Turnbaugh, P.J., Ridaura, V.K., Faith, J.J., Rey, F.E., Knight, R., and Gordon, J.I. (2009). The effect of diet on the human gut microbiome: a metagenomic analysis in humanized gnotobiotic mice. *Sci Transl Med* 1, 6ra14.

Vaishnava, S., Yamamoto, M., Severson, K.M., Ruhn, K.A., Yu, X., Koren, O., Ley, R., Wakeland, E.K., and Hooper, L.V. (2011). The antibacterial lectin RegIIIgamma promotes the spatial segregation of microbiota and host in the intestine. *Science* 334, 255–58.

Van Leuven, J.T., Meister, R.C., Simon, C., and McCutcheon, J.P. (2014). Sympatric speciation in a bacterial endosymbiont results in two genomes with the functionality of one. *Cell* 158, 1270–80.

van Rensburg, J.J., Lin, H., Gao, X., Toh, E., Fortney, K.R., Ellinger, S., Zwickl, B., et al. (2015). The human skin microbiome associates with the outcome of and is influenced by bacterial infection. *mBio* 6, e01315–15.

Venn, A.A., Loram, J.E., and Douglas, A.E. (2008). Photosynthetic symbioses in animals. *J Exp Bot* 59, 1069–80.

Versini, M., Jeandel, P.Y., Bashi, T., Bizzaro, G., Blank, M., and Shoenfeld, Y. (2015). Unraveling the Hygiene Hypothesis of helminthes and autoimmunity: origins, pathophysiology, and clinical applications. *BMC Med* 13, 81.

Vijay-Kumar, M., Aitken, J.D., Carvalho, F.A., Cullender, T.C., Mwangi, S., Srinivasan, S., Sitaraman, S.V., Knight, R., Ley, R.E., and Gewirtz, A.T. (2010). Metabolic syndrome and altered gut microbiota in mice lacking Toll-like receptor 5. *Science* 328, 228–31.

Vilcinskas, A. (2015). Pathogens as biological weapons of invasive species. *PLoS Pathog* 11, e1004714.

Wada-Katsumata, A., Zurek, L., Nalyanya, G., Roelofs, W.L., Zhang, A., and Schal, C. (2015). Gut bacteria mediate aggregation in the German cockroach. *Proc Natl Acad Sci USA* 112, 15678–83.

Wade, W.G. (2013). The oral microbiome in health and disease. *Pharmacol Res* 69, 137–43.

Waters, C.N., Zalasiewicz, J., Summerhayes, C., Barnosky, A.D., Poirier, C., Galuszka, A., Ceareta, A., et al. (2016). The Anthropocene is functionally and stratigraphically distinct from the Holocene. *Science* 351, aad2622.

Weiss, B.L., Maltz, M., and Aksoy, S. (2012). Obligate symbionts activate immune system development in the tsetse fly. *J Immunol* 188, 3395–403.

Welsh, R.M., Zaneveld, J.R., Rosales, S.M., Payet, J.P., Burkepile, D.E., and Thurber, R.V. (2016). Bacterial predation in a marine host-associated microbiome. *ISME J* 10, 1540–44.

Wernegreen, J.J. (2012). Mutualism meltdown in insects: bacteria constrain thermal adaptation. *Curr Op Microbiol* 15, 255–62.

West, S.A., El Mouden, C., and Gardner, A. (2011). Sixteen common misconceptions about the evolution of cooperation in humans. *Evol Human Behav* 32, 231–62.

Whelan, K., and Quigley, E.M. (2013). Probiotics in the management of irritable bowel syndrome and inflammatory bowel disease. *Curr Opin Gastroenterol* 29, 184–89.

Wier, A.M., Nyholm, S.V., Mandel, M.J., Massengo-Tiasse, R.P., Schaefer, A.L., Koroleva, I., Splinter-Bondurant, S., et al. (2010). Transcriptional patterns in both host and bacterium underlie a daily rhythm of anatomical and metabolic change in a beneficial symbiosis. *Proc Natl Acad Sci USA* 107, 2259–64.

Wilkinson, T.L., Ashford, D.A., Pritchard, J., and Douglas, A.E. (1997). Honeydew sugars and osmoregulation in the pea aphid *Acyrthosiphon pisum*. *J Exp Biol* 200, 2137–43.

Williams, T.A., Foster, P.G., Cox, C.J., and Embley, T.M. (2013). An archaeal origin of eukaryotes supports only two primary domains of life. *Nature* 504, 231–36.

Wilson, M.C., Mori, T., Ruckert, C., Uria, A.R., Helf, M.J., Takada, K., Gemert, C., et al. (2014). An environmental bacterial taxon with a large and distinct metabolic repertoire. *Nature* 506, 58–62.

Winter, S.E., Thiennimitr, P., Winter, M.G., Butler, B.P., Huseby, D.L., Crawford, R.W., Russell, J.M., et al. (2010). Gut inflammation provides a respiratory electron acceptor for *Salmonella*. *Nature* 467, 426–29.

Wollman, E. (1911). Sur l'elevage des mouche steriles. Contribution a la connaissance du role des microbes dans les voies digestives. *Ann Institute Pasteur* (Paris) 25, 79–88.

Wong, A.C., Chaston, J.M., and Douglas, A.E. (2013). The inconstant gut microbiota of *Drosophila* species revealed by 16S rRNA gene analysis. *ISME J* 7, 1922–32.

Wong, C.N., Dobson, A.J., and Douglas, A.E. (2014). Gut microbiota dictates the metabolic response of *Drosophila* to diet. *J Exp Biol* 217, 1894–901.

Wooding, A.L., Wingfield, M.J., Hurley, B.P., Garnas, J.R., de Groot, P., and Slippers, B. (2013). Lack of fidelity revealed in an insect-fungal mutualism after invasion. *Biol Lett* 9, 20130342.

Wu, H.J., and Wu, E. (2012). The role of gut microbiota in immune homeostasis and autoimmunity. *Gut Microbes* 3, 4–14.

Wybouw, N., Dermauw, W., Tirry, L., Stevens, C., Grbic, M., Feyereisen, R., and Van Leeuwen, T. (2014). A gene horizontally transferred from bacteria protects arthropods from host plant cyanide poisoning. *Elife* 3, e02365.

Xu, J., Bjursell, M.K., Himrod, J., Deng, S., Carmichael, L.K., Chiang, H.C., Hooper, L.V., and Gordon, J.I. (2003). A genomic view of the human-*Bacteroides thetaiotaomicron* symbiosis. *Science* 299, 2074–76.

Yamagata, N., Ichinose, T., Aso, Y., Placais, P.Y., Friedrich, A.B., Sima, R.J., Preat, T., Rubin, G.M., and Tanimoto, H. (2015). Distinct dopamine neurons mediate reward signals for short- and long-term memories. *Proc Natl Acad Sci USA* 112, 578–83.

Yano, J.M., Yu, K., Donaldson, G.P., Shastri, G.G., Ann, P., Ma, L., Nagler, C.R., Ismagilov, R.F., Mazmanian, S.K., and Hsiao, E.Y. (2015). Indigenous bacteria from the gut microbiota regulate host serotonin biosynthesis. *Cell* 161, 264–76.

Yu, D.H., Gadkari, M., Zhou, Q., Yu, S., Gao, N., Guan, Y., Schady, D., et al. (2015). Postnatal epigenetic regulation of intestinal stem cells requires DNA methylation and is guided by the microbiome. *Genome Biol* 16, 211.

Zac Stephens, W., Burns, A.R., Stagaman, K., Wong, S., Rawls, J.F., Guillemin, K., and Bohannan, B.J. (2016). The composition of the zebrafish intestinal microbial community varies across development. *ISME J* 10, 644–54.

Zaidman-Remy, A., Herve, M., Poidevin, M., Pili-Floury, S., Kim, M.S., Blanot, D., Oh, B.H., Ueda, R., Mengin-Lecreulx, D., and Lemaitre, B. (2006). The *Drosophila* amidase PGRP-LB modulates the immune response to bacterial infection. *Immunity* 24, 463–73.

Zeeuwen, P.L., Kleerebezem, M., Timmerman, H.M., and Schalkwijk, J. (2013). Microbiome and skin diseases. *Curr Opin Allergy Clin Immunol* 13, 514–20.

Zelezniak, A., Andrejev, S., Ponomarova, O., Mende, D.R., Bork, P., and Patil, K.R. (2015). Metabolic dependencies drive species co-occurrence in diverse microbial communities. *Proc Natl Acad Sci USA* 112, 6449–54.

Zhang, Q., Widmer, G. and Tzipori, S. (2013). A pig model of the human gastrointestinal tract. *Gut Microbes* 4, 193–200.

Zhao, L. (2013). The gut microbiota and obesity: from correlation to causality. *Nat Rev Microbiol* 11, 639–47.

Zheng, H., Dietrich, C., Radek, R., and Brune, A. (2015). *Endomicrobium proavitum*, the first isolate of Endomicrobia class. nov. (phylum Elusimicrobia)—an ultramicrobacterium with an unusual cell cycle that fixes nitrogen with a Group IV nitrogenase. *Environ Microbiol* 18, 191–204.

Zoetendal, E.G., Raes, J., van den Bogert, B., Arumugam, M., Booijink, C.C., Troost, F.J., Bork, P., Wels, M., de Vos, W.M., and Kleerebezem, M. (2012). The human small intestinal microbiota is driven by rapid uptake and conversion of simple carbohydrates. *ISME J* 6, 1415–26.

Zurek, L., and Ghosh, A. (2014). Insects represent a link between food animal farms and the urban environment for antibiotic resistance traits. *Appl Environ Microbiol* 80, 3562–67.

INDEX

Acanthamoeba, 24

Acetobacter pomorum, 197

Acinetobacter, 45

Acinetobacter baylyi, 16

Acropora millepora, 138, 140

Actinobacteria, 15, 39, 131, 143, 149

Actinomyces, 42

Acyrthosiphon pisum, 88, 182

addiction: in animal-microbial associations, 159–63; *Drosophila melanogaster, Wolbachia*, and *Drosophila* C virus (DCV), 160, 162; evolutionary, and consequences, 161

Aedes aegypti, 70, 71

Agaricus bisporus, 26

Akkermansia muciniphila, 46, 59, 64

Algoriphagus, 28–29

Alistipes, 43

Altered Schaedler Flora (ASF), 51

Alteromonas, 87

Amboseli Baboon Research Project, 115

Ambystoma maculatum, 192

amino acids, microbial synthesis, 6–7, 15, 168–73

Amphimedon queenslandica, 30, 32, 37

Amylostereum areolatum, 203

Amylostereum chailletii, 203

animal behavior: gut microbiota and human mental health, 110–12; host-pathogen interactions, 94; microbes and feeding behavior, 95–102; microbial effects on neural pathways, 106–7; microbial products and microbiome-gut-brain axis, 107–10; microbiome-gut-brain axis and, 102, 103; microbiome in, 10; microbiota and mental well-being, 102–12, 119; microorganisms driving, 93–95; rodent models insight into, 103–6

animal communication: fermentation hypothesis, 112–14; microbes and, 112–18, 119–20; microbial fermentation products driving social behavior of insects, 115–17; signal evolution by host capture of microbial metabolism, 117–18; social interactions and distribution of microorganisms, 114–15

animal feeding: evidence for microbial effects on, 97, 98; evolutionary scenarios for, 100–102; gut microbiota and food consumption, 98; mechanisms of microbial impacts on regulation of, 98–100; microbiota and regulatory network, 95–97, 118–19; microorganisms reinforcing behavioral traits, 101; serotonin and, 99–100

animal guts: evolutionary origin, 33–36; horizontal gene transfer among bacteria in, 174, 175;

animal immune system, 31, 92; direct effects of microbiota, 76–79; distance-based interactions with microbiota, 79, 81–83; microbiota and function of, 76–84; microorganisms and maturation of, 83–84

animal-microbial associations: addiction and consequences of, 159–63; costs and benefits of, 153–63; evolutionary consequences of, 10–11; exploitation of, 156–58, 159; reciprocity in, 153–56

animal models: *Caenorhabditis elegans*, 54–55; *Drosophila*, 53–54; genetic lesions in, 66; health consequences of microbiota loss, 62–63; in human microbiome research, 50–55; mouse, 50–52; zebrafish, 52–53

animals: in Anthropocene, 200–205; associations with microorganisms generating diversity, 180–84; description of, 1–3; determinants of animal phenotype, 197–200; developmental patterns of bacterial communities in, 146; developmental systems, 195–97; as ecosystems, 121–23; genotype and environment interaction (G X E), 197; inheritance of acquired characteristics, 199–200; mass extinction and microbiome, 204–5;

animals (*continued*): microbial dimensions of invasive, 202–4; microbial interactions and adaptivity of phenotypes, 198–99; microbiology of, 5–9, 11; phylogenetic perspective of, 192–93; physiological systems, 193–95; relationships, 29; responses to climate change, 200–201; scope of, 192–97; symbiosis-mediated speciation of, 184–89. *See also* ecological processes; inner ecosystem of animals

Anopheles, 194

Anthropocene, 200–205

antimicrobial peptides (AMPs), 69; structuring microbial community, 72, 73

Aphanomyces astaci, 202

Aphidius ervi, 88

Aphis fabae, 198

Apodemus sylvaticus, 203

apparent competition: gut bacteria and viruses, 91; microbiota and, 90–91

Archaea, 13, 14, 20

arminins 73, 74

Asobara tabida, 163

Aspergillus nidulans, 26

atherosclerosis, gut microbiota and, 57–58

Atopobium, 143

Auchenorrhyncha, 168–69

Bacteroidaceae, 166

Bacteroides, 43, 46, 99, 143; *B. fragilis*, 78; *B. ovatus*, 132; *B. plebeius*, 174; *B. thetaiotaomicron* (*B. theta*), 76, 77

Bacteroidetes, 39, 43, 148, 149

Bacteroidia, 131

Bartonella, 203

basal animals, microbiology of, 29–33

Baumannia, 15–16, 168

Bdellovibrio, 134

behavior. *See* animal behavior

Bifidobacteriaceae, 166

Bifidobacterium, 46, 63, 141, 146–47; *B. infantis*, 104, 105; *B. longum*, 110, 111

Bilophila wadsworthia, 141

Black Queen hypothesis, 87

Blatella germanica, 116–17

blood-brain barrier, 107–9

Bordetella, 23

Botryosphaeria dithidea, 184

branched chain amino acids 169, 170

Buchnera, 169, 170, 171, 179; *B. aphidicola*, 48, 49, 156–57

Burkholderia, 25; *B. cepacia*, 18–19

B vitamin synthesis, 7

byproduct mutualism, 155

Caenorhabditis elegans, as model for microbiome research, 54–55

Calliphora, 8

Callosobruchus chinensis, 180

Campylobacter, 56

Candida, 24; *C. albicans*, 24, 27

cardiovascular disease, gut microbiota and, 57–59

Carnivora, 113

Carsonella, 170–71, 180

chemical communication: in bacteria, 18

chemosynthesis, 7

chemosynthetic symbioses, 180–83

chimpanzees, gut microbiota of, 114–15

Chlamydia, 24

Chlorella, *Paramecium bursaria* and, 157–58, 159, 190

choanoflagellates, 13, 29–30; interactions between bacteria and, 27–29

Citrobacter rodentium, 76, 88–89

climate change, 200–201

clindamycin, mice treated with, 143, 144

Clostridium, 41, 46; *C. coccoides*, 60; *C. difficile*, 64, 143, 144; *C. innocuum*, 141; *C. scindens*, 64

Cnidaria, 29, 34

CO_2 fixation, 7

Comamonadaceae, 146

commensal, term, 4–5

Commensalibacter intestini, 72, 73

communication between protists and bacteria, 23–24

community state types (CSTs), bacterial, 44

coral bleaching, 137–39, 140

core microbiota 43–45

Corynebacterium, 42

Crithidia, 23

Crocuta crocuta, 113

Crohn's disease, 56

Cryptococcus fagisuga, 203

Cryptococcus neoformans, 25, 26

Curvibacter, 145

Cyanothece spp., 24

dental plaque, bacterial communities in, 41–42

Desulfobulbaceae, 146

Desulfovibrio spp., 46, 141

DIABIMMUNE project, 45

Dictyostelium discoideum, 24

discordant twins, healthy and kwashiorkor microbiota, 141–42

Dorea, 46

Drosophila, 8, 48, 49, 64, 69, 70, 84, 91, 135, 167; *Drosophila* C virus (DCV), 89, 90, 160, 162; *D. neotestacea*, 84, 85, 86; *D. paulistorum*, 189; *D. recens*, 189; *D. subquinaria*, 189; genotype variation, 197, 199; as model for microbiome research, 51, 53–54
Drosophila melanogaster, 89, 160, 162, 189; gut microbiota and mating preference of, 185, 186; mediated outbreeding of *S. cerevisiae*, 177, 178
dysbiosis: host-microbiota interactions, 4, 55, 56

ecological processes, 127–37; among-microbe interactions, 130–33, 136–37; co-occurrence patterns among microorganisms, 131; host determinants of microbial community composition, 135–36; host-microbe interactions, 136–37; predators and parasites in microbiomes, 133–34; taxonomic and functional indices of community composition, 127–30
ecosystems: animals as, 121–23; biological rhythms, 148. *See also* ecological processes; inner ecosystem of animals
embryogenesis: in animals, 195–97
endosymbioses, 4
Enterobacteriaceae, 143, 170
Enterococcus, 143, 144
Entomophaga glylli, 93
Entotheonella, 87
epidemiology, studies of human microbiome, 45–46
EPS (extracellular polymeric substances), 17
Erwinia toletana, 19
Escherichia coli, 16, 24, 54, 56, 71, 88, 134, 167, 176
essential amino acid synthesis, 7, 153–55, 169–73
Eubacteria, 14
Eubacterium rectale, 43, 60
eukaryotes, 8; ancestors of genome, 20; choanoflagellates and bacteria, 27–29; diversity of, 14; fungi and bacteria, 24–27; interactions with bacteria, 36–37; microorganismal origins of, 20–23; protists and bacteria, 23–24; symbiotic origin of modern, 20–23; ubiquity of microbial associations in, 23–29
Eumetazoa: polarized epithelium and gut of, 34
Euplotidium, 23
Euprymna scolopes, *Vibrio fischeri* and, 73, 79, 80, 125–26, 149

European Metagenomics of the Human Intestinal Tract (MetaHIT), 38, 43
Euscelis incisus, 195
evolution: animal feeding behavior and microbial impact, 100–102; evolutionary history of life, 14; metabolic coevolution of insect-bacterial symbioses, 168–73; microbiome science, 152–53, 191–92
evolutionary specialization: congruent host-microbial phylogenies, 164–65; genome reduction and coevolution, 167–73; phylogenetic patterns and interpretation, 164–67; strains of gut bacterium *Lactobacillus reuteri*, 165–66; symbiosis for diversification, 173–89. *See also* symbiosis
exploitation: of animal-microbial associations, 156–58, 159; of *Chlorella* symbionts by *Paramecium bursaria*, 159
exploitation competition, microbiota and, 88–89, 90
extracellular matrix, 17

Faecalibacterium spp., 46; *F. prausnitzii*, 57, 64
fecal microbiota transplantation (FMT), 63–64
feeding. *See* animal feeding
fermentation hypothesis 112–14, 117–18
Firmicutes, 15, 39, 41, 43, 46, 113, 148, 149
Flavobacteria, 131, 145, 146
fungi, interacting with bacteria, 24–27
Fusarium, 84
Fusobacteria, 39, 131
Fusobacterium nucleatum, 57

Gardnerella, 143
genomes, reduction and coevolution, 21–23, 167–73
genotype and environment interaction (G X E), 197
Glossina, 83
Gluconobacter morbifer, 72, 73
gut barrier, 107–9
gut-brain axis, 102–8, 197
gut microbiota: cardiovascular disease and, 56–59; of chimpanzees in Tanzania, 114–15; diet and microbial diversity of, 60–62; gut disease and, 56–57; humans, 165–66; long-distance immune system effects on, 79, 81–83; metabolic disease and, 59–60

Haemophilus influenzae, 90
Halobacteriovorax, 134

Halomonadaceae, 170
Hamiltonella, 79, 80, 88, 182–83, 200
Hartmannella, 24
Helicobacteraceae, 145, 146
Helicobacter pylori, 57, 204
Hemiptera, 168–69
Herpestes auropunctatus, 112
Histidine, 155–56, 172
Hodgkinia, 168, 172
holobiont, 5
homeostasis, host-microbiota interactions, 55, 56
Howardula, 84, 85, 86–87
human disease, microbiota and, 55–60
human health: correlation, causation, and mechanism of, 46–49; gut microbiota and human mental health, 110–12; microbiome and, 38–39, 64–65; microbiota loss and, 62; studying microbiota effects, 45–55
human microbiome, 9–10: abundance of bacteria in human body, 40; among-individual variation in, 43–45; animal models in research of, 50–55; bacterial communities in dental plaque, 41–42; biogeography of, 39–45; cardiovascular disease and, 57–59; core microbiota of, 43–45; correlation, causation, and mechanism, 46–49; distribution and abundance of microorganisms, 39–42; epidemiological studies of, 45–46; gut microbiota and human health/disease, 57–60; homeostasis and dysbiosis, 55, 56; metabolic syndrome and, 59–60; microbiota of human skin, 42; microorganisms in colon, 41–42; protective mucus in gut, 40–41; studying gut microbiota, 165–66
Human Microbiome Project (HMP), 38, 43
Hyaena hyaena, 113
Hydra, 34, 145, 158; *Chlorella* in cells of, 192; *Fusarium* and, 84; *H. oligactis*, 72, 74; *H. viridissima*, 72; *H. vulgaris*, 72, 74, 145; microbiota in, 74
Hypothenemus hampei, 179

immune effectors, evolution as positive regulators, 74–75
immune system: abundance and distribution of microorganisms, 68–71; of animal, 31, 92; animal models of, 66; dictating nutrient supply to microbiota, 75–76; engagement with microbiota, 67; interference competition, 86–88; protective mechanisms of microbiota, 84, 85; provisioning of sugar fucose, 77;

regulation of composition of microbiota, 72–75; regulation of microbiota, 68–76; understanding the, 66–67. *See also* animal immune system
immunoglobulin A (IgA), role in gut microorganisms, 68–69
inflammatory bowel disease (IBD), 56–57
info-chemicals 16–19
inheritance of acquired characteristics, 199–200
inner ecosystem of animals: biological rhythms, 148; bleaching susceptibility of coral, 137–39, 140; developmental patterns of bacterial communities in animals, 146; diversity, stability and alternative stable states of, 138, 140–44; functions of, 137–50; host rhythms and function of, 147–50; primary succession in, 144–47; properties of simple animal microbiome, 137–40
insects: metabolic coevolution in insect-bacterial symbioses, 168–73; microbial fermentation products driving social behavior of, 115–17
interference competition, microbiota and, 86–88
invasive animals, microbial dimensions of, 202–4
irritable bowel syndrome (IBS), 46

Kiwa puravida, 145, 146
Klebsiella, 25, 26, 56
kwashiorkor microbiome, 141–42
kynurenines, 108–9

Lachnospiraceae, 166
Lactobacillus, 41, 45, 59–60, 62, 63, 99, 141, 143; *L. helveticus*, 110; *L. plantarum*, 185, 186; *L. reuteri*, 165–66, 168, 190; *L. rhamnosus*, 106–7
Lagenidium callinectes, 87
Larrea tridentata, 198
Legionella pneumophila, 24
Listeria monocytogenes, 81
Lokiarchaeota 14, 20
Longitudinal Study of Australian Children (LSAC) program, 45
luminescence, 7

Magicicada tredecim, 172
Malassezia globosa, 42
Malassezia restricta, 42
mass extinction event, 60–64; health consequences of microbiota loss, 62–63;

microbial communities of low diversity, 60–62; microbiome and, 204–5; restoration of microbiota, 63–64

Medicago, 182

mental well-being: gut microbiota and human, 110–12; microbial metabolites modulating gut-brain axis, 107–10; microbiome-gut-brain axis and, 102, 103; rodent models studying effects of microbiota, 103–6. *See also* animal behavior

metabolic coevolution of insect-bacterial symbioses, 168–73

metabolic disease, gut microbiota and, 59–60

metabolomics, 44

metagenomics, 44

Methylococcaceae, 145, 146

Methylophilaceae, 146

mice: antibiotic perturbation of cecal microbiota of, 144; gut microbiota and behavioral traits of, 104, 105; gut microbiota and food consumption, 97, 98; gut microbiota and obesity in, 47, 48; microorganisms and host immune function of, 78; as model for microbiome research, 50–52; models studying microbiota impact on behavior of, 103–6; production of trimethylamine (TMA) in, 117–18; type 1 diabetes (T1D) research, 81–82

microbe-assisted molecular patterns (MAMPs), 31, 32, 109

microbial associations, ubiquity in eukaryotes, 23–29

microbial fermentation products: driving social behavior of insects, 115–17; as group identity signals, 114–15

microbiome, 8; biogeography of human, 39–45; evolution and science of, 152–53, 191–92; human health and, 38–39, 64–65; mass extinction and, 204–5; predators and parasites in animal, 133–34; role in shaping animal behavior, 10; term, 3. *See also* human microbiome

microbiome-gut-brain axis: microbial products underpinning, 107–10; microbiota-mediated regulation of, 102, 103

microbiota: animal immune function and, 76–84; apparent competition, 90–91; engagement with immune system, 67; exploitation competition, 88–89, 90; gut microbiota and gut disease, 56–57; human disease and, 55–60; immune effectors and regulation of, 68–76; immune system dictating nutrient availability,

75–76; immunological regulation of composition of, 72–75; interference competition, 86–88; protective mechanisms of, 84, 85; restoration of the, 63–64; as second immune system, 84–91; term, 3. *See also* animal immune system

microorganisms: abundance and distribution of animal-associated, 123–27; animal-associated, with limited or no free-living populations, 123–26; animal-associated, with substantial free-living populations, 126–27; communities of low diversity, 60–62; co-occurrence patterns among, 131; distribution and abundance in human microbiome, 39–42; distribution between animal host and external environment, 125; driving animal behavior, 93–95; immunological control of abundance and distribution of, 68–71; interactions with animal host, 6; interactions with gut epithelium, 35–36; maturation of immune system and, 83–84; as source of genetic novelty, 177, 179–80; study of, 2. *See also* inner ecosystem of animals

mitochondria: candidate bacterial-derived organelles, 22; eukaryotes and microorganisms, 21–22

Montastraea annularis, 137, 140

mouse. *See* mice

Mucus, 40–41, 68–69, 73, 92, 129, 134

Mycobacterium, 24; *M. leprae*, 168

Myodes (=*Clethrionomys*) *glareolus*, 203

Nasonia, 186, 187; *N. giraulti*, 186, 187; *N. vitripennis*, 186, 187

Neisseriaceae, 42

Neisseria meningitides, 34

Neonectria, 203

Neotoma lepida, 198

nitrogen fixation, 7

Oophila, 192

oxidative phosphorylation, 21

Palaemon macrodactylus, 87

Paramecium bursaria, *Chlorella* and, 157–58, 159, 190

pathogens: driving animal behavior, 93–95; host-pathogen interactions, 94

pattern recognition receptors (PRRs), 30, 32

Paulinella, 23

Peromyscus, 167

phagocytosis, 31

phenotypes, 2

photosynthesis, 7
Plasmodium, 194
polarized epithelium: animal gut and, 33–36; gut of Eumetazoa and, 34
Polistes dominula, 177
Porifera, 29
Porphyromonas, 42
Portiera, 169, 170, 171
Prevotella, 43, 143; *P. copri*, 82
primary metabolites, sharing, 15–16
Prochloron, 87
Propionibacterium acnes, 42
protective mechanisms: microbiota and, 84, 85
Proteobacteria, 15, 39, 41, 131
protists, communication between bacteria and, 23–24
Providencia sp., 187
Pseudomonas: *P. aeruginosa*, 18–19, 27; *P. putida*, 26, 27; *P. savastanoi*, 19

quorum sensing (QS) molecules, 17–19

Raffaelea lauricola, 202–3
reciprocity: in animal-microbial associations, 153–56; freeloaders in, 154, 156; overconsumers in, 154, 156
Regiella, 79, 80, 182; *R. insecticola*, 182–83, 200
Rhagoletis pomonella, 184
Rhizopus microsporus, 25
Rhodobacteraceae, 146
Rickettsia, 8, 20–22, 36, 179
Rikenellaceae, 62
Roseburia, 43
Ruminococcaceae, 166
Ruminococcus, 46

Saccharomyces cerevisiae, 177, 178
Saccharomyces paradoxus, 177
Salmonella, 84, 90–91, 141, 176
Salpingoeca rosetta, 28–29
Santia, 84
Saprospiraceae, 146
secondary metabolites, fungal-bacterial associations, 26
serotonin: gut microbiota and production of, 195; regulating feeding behavior, 99–100, 107
Shelfordella lateralis, 55
Shewanella, 52–53, 134
Sirex: fungal symbionts of, 203; *S. nigricornis*/*Amylostereum chailletii*, 203; *S. noctilio*/*A. areolatum*, 203

Sitophilus, 71
Sodalis pierantonius, 71
Spiroplasma, 84, 85, 86
sponges, microbiology of basal animals, 29–33
Staphylococcus, 45; *S. aureus*, 81, 134; *S. epidermidis*, 42
Sternorrhyncha, 168–69
Streptococcus, 41, 42; *S. mutans*, 42; *S. pneumoniae*, 90
Streptomyces, 88; *S. rapamycinicus*, 26
Sulcia, 15–16
Symbiodinium, 137, 139
symbioses: chemosynthetic, 180–83; diversification of microbial partners, 173–77; insect-mediated outbreeding of yeast *S. cerevisiae*, 177, 178; mediated speciation of animals, 184–89; microorganisms as source of genetic novelty, 177, 179–80; microorganisms generating animal diversification, 180–84; term, 3–4
Synechococcus, 84
syntrophy, 15–16

Tenericutes, 148
Tettigades undata, 172
Thiotrichaceae, 145, 146
toxin degradation, 7
toxin synthesis, 7
Toxoplasma gondii, 93
Tremblaya, 169, 170, 171
Trichonympha, 24
Trifolium, 182
type 1 diabetes (T1D), 45–46, 81–82
type 2 diabetes (T2D), 58

U.S. Human Microbiome Project (HMP), 38, 43

Veillonella, 41
Verrucomicrobia, 39, 148
Verrucomicrobiaceae, 146
Vibrio, 52–53; *V. fischeri* and *Euprymna scolopes*, 73, 79, 80, 125–26, 149

Wigglesworthia, 80, 83–84
Wolbachia, 89, 90, 160–63, 180, 187–89

Xyleborus glabratus, 202–3

zebrafish: gut microbiota of, 145; as model for microbiome research, 51, 52–53

A NOTE ON THE TYPE

This book has been composed in Adobe Text and Gotham. Adobe Text, designed by Robert Slimbach for Adobe, bridges the gap between fifteenth- and sixteenth-century calligraphic and eighteenth-century Modern styles. Gotham, inspired by New York street signs, was designed by Tobias Frere-Jones for Hoefler & Co.